Genetic Engineering

Genetic Engineering

A Primer

Walter E. Hill

The University of Montana
Missoula, Montana, USA

CRC Press
Taylor & Francis Group
Boca Raton London New York

CRC Press is an imprint of the
Taylor & Francis Group, an **informa** business

A TAYLOR & FRANCIS BOOK

First published 2000 by Overseas Publishers Association

This edition first published 2002
by Taylor & Francis

Published 2019 by CRC Press
Taylor & Francis Group
6000 Broken Sound Parkway NW, Suite 300
Boca Raton, FL 33487-2742

First issued in paperback 2019

No claim to original U.S. Government works

ISBN 13: 978-0-367-45490-6 (pbk)
ISBN 13: 978-0-415-30007-0 (hbk)

Visit the Taylor & Francis Web site at
http://www.taylorandfrancis.com

and the CRC Press Web site at
http://www.crcpress.com

Every effort has been made to ensure that the advice and information in this book is true and accurate at the time of going to press. However, neither the publisher nor the authors can accept any legal responsibility or liability for any errors or omissions that may be made. In the case of drug administration, any medical procedure or the use of technical equipment mentioned within this book, you are strongly advised to consult the manufacturer's guidelines.

British Library Cataloguing in Publication Data
A catalogue record for this book is available from the British Library

Front Cover: Figure 3.8 DNA Replication

To

my wife *Annette*
and
my children

CONTENTS

PREFACE

I wrote this book to help you appreciate the tremendous power and potential that is at our doorstep in the area of genetic engineering. I have started with some very basic concepts so that we can keep on the same footing throughout the book. My idea is to share my awe and delight in the tremendous advances that have been made in the biological sciences, which now allow us to transform living things.

We have been shown fictitious accounts in books, movies, and television in which scientists have done some bizarre things. Can we really recreate dinosaurs from their ancestral blood? Can we clone replicas of famous people? Can we make tomatoes that can be picked ripe, but not soften on the grocery shelves? Can we cure genetic diseases by gene therapy? What are the potential hazards involved in these possibilities? These and similar questions are issues that we want to deal with in this book.

I purposely designed this book to help those with little scientific background become conversant with the area generally called *genetic engineering*—the changing of the genetic information of living organisms by design. To understand genetic engineering, we need to understand some of the living processes and the natural changes that can and do take place.

Just a little over 50 years ago, DNA was found to be the carrier of genetic information. Before long it was understood how the information was stored, coded, transferred, and translated into living things. All the information necessary to make an individual is contained in the DNA found in the original fertilized cell.

However, many mechanisms exist by which portions of the DNA are turned on and off at specific times. This regulation controls the expression of genetic information and is critical to our well-being.

So we must learn more about DNA and the genes within it to find out how to modify it. We start by looking at the chemical makeup of genes and proteins and what makes them different or similar. We then study how genes are made and how proteins are made. Then, we are ready to learn how changes can be introduced into the genes.

By making changes in the gene, important changes can then be made in the plant or animal. But how can we alter genes? Genes are too small to see, except under the most powerful magnification. How do we cut them and insert different information into them. How do we make sure that this information is used at the right time later on? How can we ensure that faulty changes don't occur? The overriding challenge in modern genetic engineering is to solve these problems. In this way, it is hoped that some of the genetic malfunctions that hurt mankind can be reversed.

There are definite limits to the kind of changes that can be made...and where they can be made. We will discuss these in light of scientific limitations as well as ethical and moral implications. We will also look at the potential for the future and the impact that all of these genetic engineering changes will have on our lives.

I have placed key terms in bold-faced type to emphasize them. The glossary defines these terms more fully and allows you to define more clearly some of these unusual terms. I have tried to include only that which is essential so as not to drown you...only quench your thirst.

ACKNOWLEDGMENTS

I am indebted to a whole host of individuals, who have trained and nurtured me in the ways of science. To both professors and graduate students, I give my thanks. And special thanks to my editors, Alison Kelley and Sally Cheney, who have been patient and helpful beyond measure.

I
BUILDING BLOCKS OF LIVING THINGS

1

MATTER AND LIVING THINGS

WHAT YOU WILL LEARN IN THIS CHAPTER

- The attributes of a cell
- What living things are composed of
- How elements and molecules are bound together
- The kinds of biological molecules

It takes little more than intuition for us to tell when something is alive; yet, trying to define what life is can get very difficult.

Is a tree alive?
Is a rock alive?
Is water alive?
Is a cell alive?

Let's take a living cell apart and look at each part.

Is the nucleus alive?
Is the DNA [deoxyribonucleic acid] alive?
Is the cell membrane alive?

So, what is life? Every substance is made up of the basic elements that are present around us. Yet, clearly, elements are not alive, nor are many of the substances that they make.

Although it is difficult to define living things exactly, we can list characteristics that are found in all living things. The list may be quite long, but for our purposes, we will use four main characteristics to define living things.

Characteristics of Living Organisms
1. Have structure or shape 2. Can make and break down molecules 3. Can transform energy 4. Can reproduce themselves

LIVING THINGS

To be defined as a living organism, the organism must meet all four of the following criteria (see box).

1. Living organisms have **structure** or shape. Many of the structural features are due to special proteins that have been made by the cells for that very purpose.
2. Living organisms can **make and break down complicated molecules**. This process is often directed by specialized proteins called **enzymes**.
3. Living organisms **transform energy** from one form to another. They receive energy in the form of sunlight, food, and heat and change it into other forms, such as chemical energy or motion, which are more useful to the organism.
4. Living things **reproduce themselves**. This means that the characteristics of the parents can be transferred to their children—and that the children or offspring can eventually reproduce themselves as well. In some cases, as with the tadpole and the frog and with the caterpillar and the moth, major transformations happen along the way.

All things on earth are made up of the more than 100 elements, as shown in Figure 1-1. Many elements are rare, but some are common. The earth itself has much oxygen, silicon, aluminum, iron, and calcium and lesser amounts of many other elements. On the other hand, living things of the earth are composed mainly of carbon, nitrogen, oxygen, hydrogen, and a few other elements.

Cells are the smallest units contained in all living things. They come in various forms and kinds. Some cells are loners and exist only as a single and complete entity. Other cells are social and

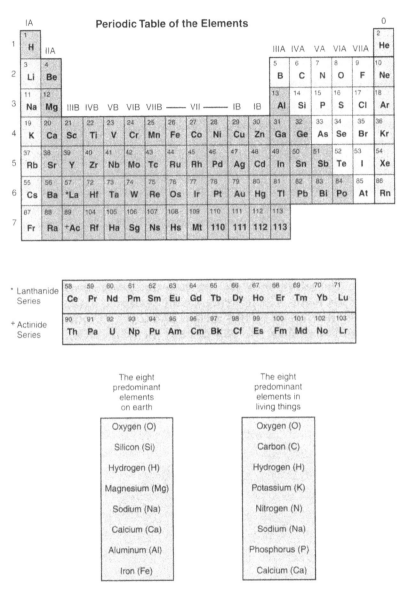

Figure 1-1   Chart showing periodic table of the elements, eight predominant elements of the earth, and eight predominant elements of living things.

can live only if they coexist with more of their kind. Some cells have specialized functions. For instance, the function of some cells is to make hair; these cells use much of their energy and many cellular mechanisms primarily to serve that one function.

Other cells make muscle protein, hemoglobin, fingernails, and
so on. Each cell follows a defined plan. And most cells also have
to make other cells like themselves.

Multicelled Organisms

Larger organisms are composed of more complicated cells,
which are often organized in specific ways. Figure 1-2 is a
diagram of a typical cell from a higher organism. Note that
whole areas of specialization can be seen within the cell. These
specialized areas of cells are called **organelles**, which can be
thought of as specific rooms in a manufacturing plant. For
instance, the **nucleus** of a cell is the place where the genetic
plans (in the form of deoxyribonucleic acid [DNA]) are kept and
distributed. This is equivalent to the executive office suite in a
manufacturing plant. The **mitochondria** are the furnace and
power rooms where energy is made and distributed throughout
the cell. The **ribosomes** are the manufacturing rooms, where
proteins are made. The proteins are the products, made by using
the plans distributed by the nucleus and using the energy made
by the mitochondria. We will discuss the ribosomes and their
functions in more detail in Chapter 4.

All the information necessary for each cell to "do its own
thing," as well as all other things the cell needs, is always

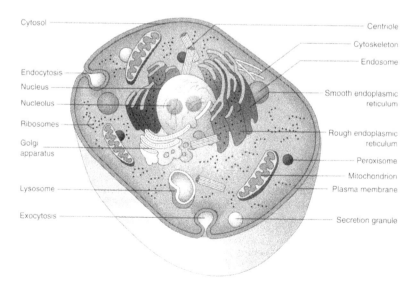

Figure 1-2   A typical cell from a higher organism.

present in the DNA. But some of it is not used until the right time. So the DNA must contain two separate kinds of information:

- Information dealing with *what* is to be made. This information must contain detailed instructions for making each of the many thousands of different kinds of proteins that are present in an organism.
- Information needed to tell the cell *when* to make these proteins. Some kind of an off/on (or dimmer) switch is essential, which the cell can use to regulate the amount of a given type of protein to be made.

The various cells from larger organisms, all functioning in their own specialized way, initially came from one fertilized egg cell (Fig. 1-3).

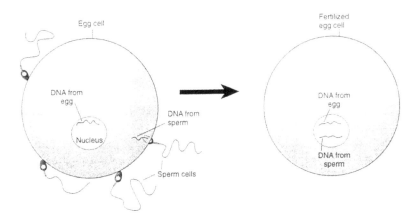

Figure 1-3. Fertilization of an egg cell. Sperm cells attach to the egg cell, and one sperm is allowed to donate its genetic material (DNA) to the egg cell. The egg cell is then fertilized and immediately beginning to grow and make new cells genetically identical with itself.

Single-Celled Organisms

Many organisms are just single cells. Among the single-celled organisms are **bacteria**. Bacteria live everywhere around us, yet are invisible except under a microscope with high magnification. Each tiny bacterium is roughly 10% of the size of cells from higher organisms; yet, each is certainly alive (Fig. 1-4).

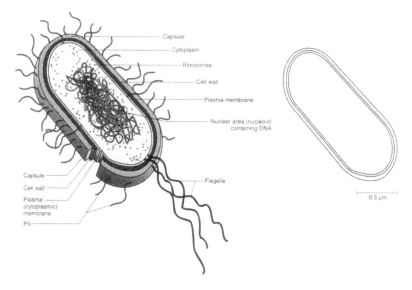

Figure 1-4   Typical bacterial cell. Note that this structure is much less complex than that found in cells from higher organisms (see Fig. 1-2). Still, bacteria maintain all of life's essential functions.

Many processes that bacteria use to survive are the same as those used in higher organisms, except that they are simpler. Scientists have studied bacteria, since living processes can be studied in detail in these cells and can shed much light on the way these processes occur in us and other larger organisms. In addition, bacteria are especially useful as tools in genetic engineering because their DNA can be manipulated easily.

ELEMENTS AND MOLECULES

If we were to break down the simplest bacterium into various component parts and then break those parts into fundamental units, we would find that the bacterium is composed of **atoms** (the fundamental unit of every element) of various elements that are bound and grouped together in specific ways. The atoms of each different element have different numbers of protons and electrons. **Protons** are particles containing a positive charge (+) and are found in the nucleus of an atom (neutrons, which contain no charge, are also found here). **Electrons** are much smaller, have a negative charge, and circulate around the positively charged nucleus, much like satellites around our planet (Fig. 1-5).

Certain atoms like to group together. For instance, hydrogen and oxygen are elements seldom found as single atoms, but are commonly found in air as twins:

$$H_2 \text{ and } O_2 \text{ or } H\text{-}H \text{ and } O\text{-}O$$

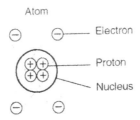

Figure 1-5 Typical structure of an atom showing the nucleus, protons, and electrons.

Two or more atoms coupled together form a **molecule**. The bonds that hold them together are shown as "-" between the atoms. Molecules composed of more than one element are called **compounds**. But elements, molecules, and compounds are not themselves living entities. They are just building blocks bonded together in many ways to give a variety of products. For instance, two hydrogen atoms and a single oxygen atom bond together to form water.

Water

Let's look at water more closely. It is simple in structure, but has fascinating properties. Hydrogen is flammable and oxygen sustains combustion, but bonded together they make water, a fire retardant! So it is with many compounds: When bonded together, they often become substances that have far different characteristics from those of the atoms or molecules of which they are made (Fig. 1-6).

Water also has many unusual properties when compared with other liquids to which it should be similar. For instance, when water freezes, it takes up more space, which in turn makes it float (as ice). Water boils and freezes at a higher temperature

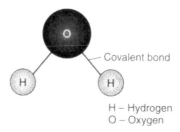

H – Hydrogen
O – Oxygen

Figure 1-6 A water molecule showing how hydrogen bonds to the oxygen. The hydrogen and oxygen atoms never line up, but are always at defined angles to each other

than other similar compounds (such as methane and ammonia). Perhaps most of all, water produces hydrogen-bonds (a very weak bond) with many other compounds, making water a great solvent. These unusual characteristics make water an ideal compound for living things (Table 1-1).

Table 1.1 Unusual Properties of Water

Compound	Molecular Weight	Melting Point ($^\circ$C)	Boiling Point ($^\circ$C)	Heat of Vaporization (kJ/mol)
CH_4	16.04	−182	−162	8.16
NH_3	17.03	−78	−33	23.26
H_2O	18.02	0	+100	40.71
H_2S	34.08	−86	−61	18.66

BONDS

Bonds are the "glue" that holds together the atoms from various elements to give all matter substance and structure. If we were to analyze everything around us in atomic detail, we would find wonderful molecules and compounds with intricate lattice structures composed of atoms. These structures and their chemical makeup give all material things a variety of sizes shapes, and other characteristics. Within these structures are several types of bonds with different functions. We will discuss these important bond types (covalent, hydrogen, and ionic) as well as a special interaction of great use in living systems (hydrophobic interaction).

Covalent Bonds

In water, hydrogen and oxygen atoms are tied together in what is termed a **covalent bond**. Covalent bonds are very strong, are found almost everywhere, and literally hold things together, giving significant strength to all structures. One of the significant properties of the four most common elements of living things—carbon, oxygen, hydrogen, and nitrogen—is a great tendency to form covalent bonds (Fig. 1-7).

Covalent bonds are often called the "backbone" of molecules because they are responsible for the primary organization of the molecules. They are made when adjacent atoms share electrons, which allows the two nuclei to be bound closer together.

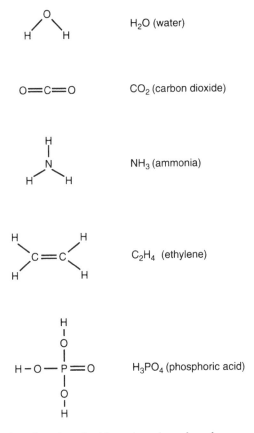

Figure 1-7   Covalent bonds. Note that these bonds can make various angles between the atoms, depending on which atoms are used.

Hydrogen-Bonds

Another kind of bond plays a critical role in living systems—the **hydrogen-bond** (H-bond). H-bonds are formed between a hydrogen atom and nearby oxygen, nitrogen, and sometimes sulfur. These bonds are relatively weak and have about 10% of the strength of covalent bonds. But what they lack in strength, they make up for in number (Fig. 1-8).

Water tends to form numerous H-bonds with other water molecules. These bonds often do not last very long, being broken and made quite rapidly. Generally, in water, about 85% of the water molecules are hydrogen-bonded to neighbors. Breaking weak H-bonds *between* water molecules does *not* break strong covalent bonds *within* the molecules. For instance, when ice is formed, even more H-bonds are made, making a solid structure of immense strength (Fig. 1-9).

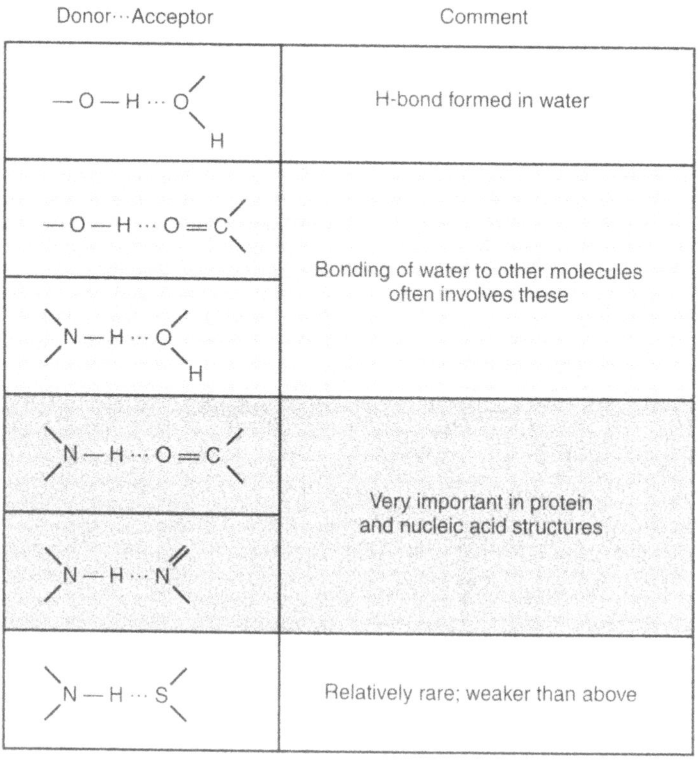

Donor···Acceptor	Comment
—O—H···O (with H below)	H-bond formed in water
—O—H···O=C	Bonding of water to other molecules often involves these
N—H···O (with H below)	
N—H···O=C	Very important in protein and nucleic acid structures
N—H···N	
N—H···S	Relatively rare; weaker than above

Figure 1-8. Various kinds of hydrogen bonds (H-bonds).

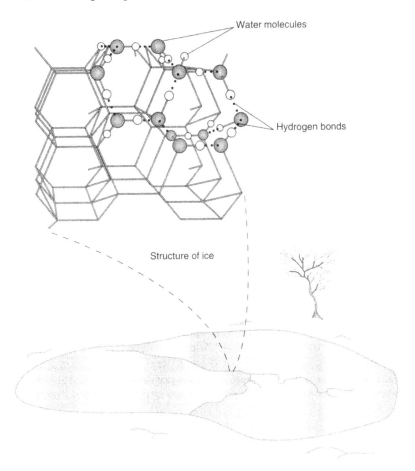

Figure 1-9 The structure of ice as it might occur in the frozen surface of a pond or river.

Water can also hydrogen-bond to other molecules, because it contains both oxygen and hydrogen, both of which are used in H-bonds. The great tendency of water to hydrogen-bond to almost everything gives it many of the remarkable characteristics.

H-bonds are also found between nonwater molecules and play a significant role in living systems, as we shall see shortly. H-bonds are important, not because of their exceptional strength, but because they do so many different things and they are so plentiful. In addition, H-bonds can be made and broken easily, which allows a great variety of structures in living things.

Ionic Bonds

Another kind of bond is formed by the electrical attraction between some molecules. Some atoms prefer to lose or gain negatively charged electrons, making themselves **ions**. By adding an electron, the atom becomes negatively charged. Losing an electron gives a net positive charge to the atom. The electrical charges between oppositely charged ions attract each other and form a bond, called an **ionic bond**. For instance, a sodium ion (Na^+) and a chloride ion (Cl^-) readily form sodium chloride (NaCl), which is common table salt. The strong ionic bond holds the sodium and chloride ions together, which generates the new compound. These ionic bonds are found in many chemical compounds and are relatively strong forces in holding things together. Ionic bonds are found in biological molecules as well, but are much less common than H-bonds.

Hydrophobic Interactions

Another interaction between atoms and molecules is very important and is really not a chemical bond at all; still, it plays a central role in living things. Some molecules don't form H-bonds. We recognize some of these as fats and oils, compounds that don't mix well with water. These portions of molecules have a tendency to avoid water molecules completely. To do so, they group together. This avoidance of water is called a **hydrophobic** (water-hating) interaction. H-bonds between solvent water molecules and hydrophobic interactions between oil molecules are the reason why "water and oil don't mix." Hydrophobic interactions are mainstays in biological structures (discussed in more detail in Chapter 2).

BIOLOGICAL MACROMOLECULES

In living things, large, complicated structures (**macromolecules**) are made from smaller molecules. For instance, plants use carbon dioxide (CO_2) and water to form carbohydrates (sugars). These sugar molecules are formed from carbon atoms combined with hydrogen and hydroxide (OH) attached. A five-carbon sugar would have the structure shown in Figure 1-10a.

But the structure does not remain a linear string of carbons. Instead, the carbons form a ring-like structure, as shown in

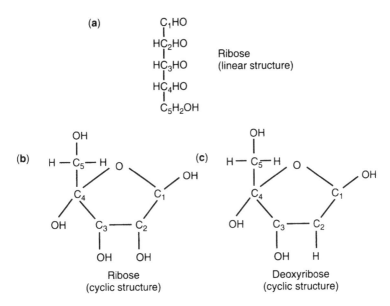

Figure 1-10   (**a**) The linear structure of ribose. (**b**) The cyclic structure of ribose and deoxyribose. These are both five-carbon sugars commonly found in genetic material (RNA and DNA, respectively). Nature favors the cyclic structure more than 99% of the time for these sugars

Figure 1-10*b*. This sugar molecule, **ribose,** is a five-carbon sugar that plays a central role in both DNA and RNA (ribonucleic acid). As a matter of fact, RNA actually stands for ribo(se)nucleic acid, and DNA stands for deoxyribo(se)nucleic acid. The only difference between ribose and deoxyribose is that ribose has an extra oxygen, as shown in Figure 1-10*b*.

These ribose (or deoxyribose) units, can be bound to additional cyclic molecules that have backbones made of both carbons and nitrogens. These molecules are called **bases**. Their structures are shown in Figure 1-11. Bases can be joined to ribose (or deoxyribose) sugar, and when a phosphate group is added, we have **nucleotides** (Fig. 1-12).

The five bases shown in Figure 1-11 are the fundamental building blocks of both DNA and RNA. They are labeled A (adenine), G (guanine), C (cytosine), and T (thymine), or U (uracil) after the first letter of their name. The first four bases, A, G, C, and T, when attached to deoxyribose sugars, are found in DNA and are called **deoxyribonucleotides** (Fig 1-12). A, G, C,

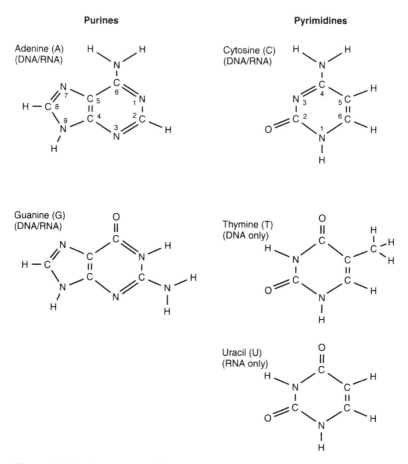

Figure 1-11 Structures of purine and pyrimidine bases found in DNA and RNA. Note that each of the atoms in the cyclic structures is numbered. In Figure 1-10, the carbons in the sugars were also numbered. When they are put together into nucleotides (see Fig. 1-12), the sugar numbers are given a prime (′) to distinguish them from the numbers in the bases.

and U when attached to ribose sugars are found in RNA and are called **ribonucleotides** (Fig 1-12). DNA and RNA are nucleic acids that are made of long chains of such nucleotides. These nucleic acid macromolecules are critically important to all living organisms and will be discussed in much more detail in Chapter 3.

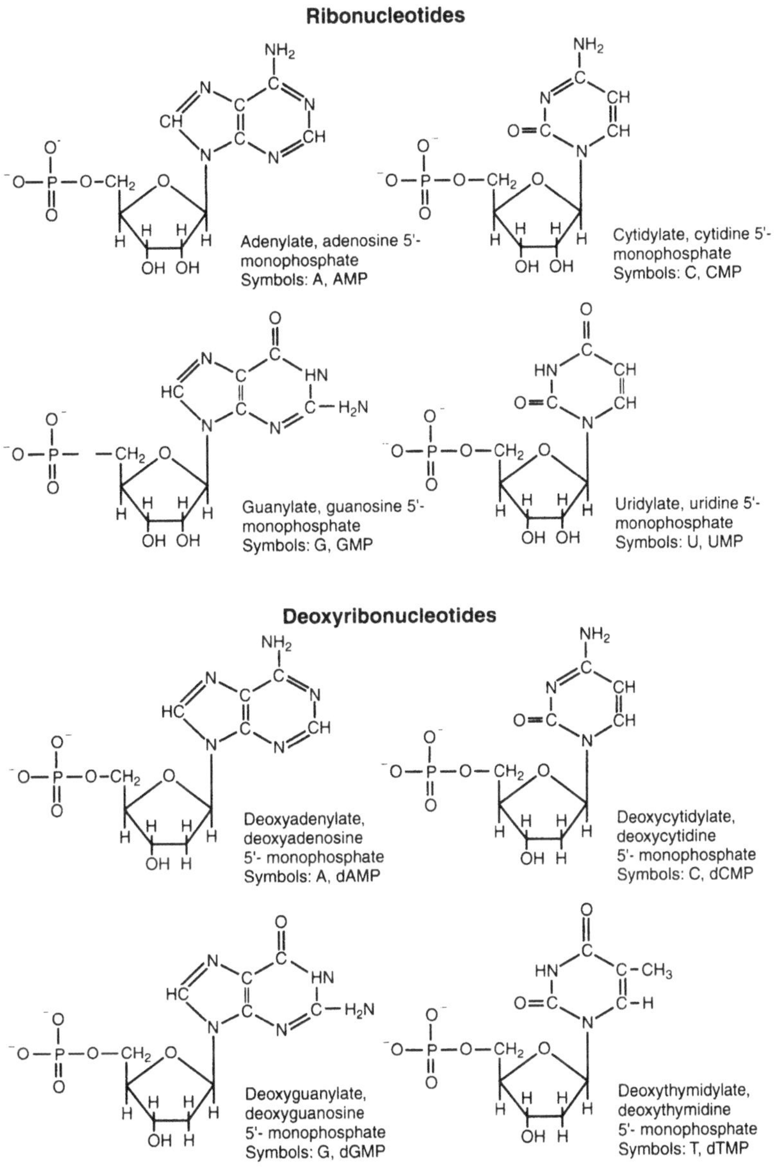

Ribonucleotides

Adenylate, adenosine 5'-
monophosphate
Symbols: A, AMP

Cytidylate, cytidine 5'-
monophosphate
Symbols: C, CMP

Guanylate, guanosine 5'-
monophosphate
Symbols: G, GMP

Uridylate, uridine 5'-
monophosphate
Symbols: U, UMP

Deoxyribonucleotides

Deoxyadenylate,
deoxyadenosine
5'- monophosphate
Symbols: A, dAMP

Deoxycytidylate,
deoxycytidine
5'- monophosphate
Symbols: C, dCMP

Deoxyguanylate,
deoxyguanosine
5'- monophosphate
Symbols: G, dGMP

Deoxythymidylate,
deoxythymidine
5'- monophosphate
Symbols: T, dTMP

Figure 1-12 Structures of the nucleotides used in RNA (ribonucleotides) and in DNA (deoxyribonucleotides). Note that the bases on
the first three across are identical. Only the lack of oxygen on the 2′
carbon of the sugar of the deoxyribonucleotides makes them different.

SUMMARY

Living things are complex structures of various elements bonded together in specific ways. Cells are the smallest units of living things and contain highly specialized organelles to allow living processes to take place. Cells contain many complex structures called macromolecules, which are formed from other molecules and elements. These are held together with various bonds, such as covalent, hydrogen, and ionic bonds. One of these complex macromolecules is deoxyribonucleic acid (DNA).

2

PROTEINS

WHAT YOU WILL LEARN IN THIS CHAPTER

- The makeup of a protein
- How proteins are put together
- How proteins provide structure to living things
- How chemical reactions use proteins to make them go

Now that we have learned how atoms and molecules are held together chemically, we can turn to the larger complexes of molecules (macromolecules) used in living systems. We used nucleic acids in Chapter 1 as an example of the complex structures that can be built from relatively simple compounds. Now, we introduce proteins as the next major category of macromolecules.

Proteins are the final product of almost all the genetic information carried in a cell. As we will see in Chapter 3, genetic material encodes all the information needed to make the structure of the proteins to be exactly what they need to be so that they can perform their many functions.

Proteins are amazingly diverse, both in structure and function. Proteins come in thousands of different shapes and forms. They are more carefully designed and built than the finest mansions. Their intricate design and form allow them to do their particular job and no other. Proteins are patterned to allow them to work effectively and efficiently, using very little energy.

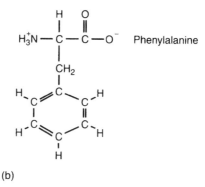

(a) Amino acid (ionic form)

(b)

Figure 2-1   Amino acids—the building blocks of proteins. (**a**) General structure of an amino acid. **R** = any of the 20 different side chains. (**b**) Three of the 20 amino acids commonly used in proteins.

AMINO ACIDS: PROTEIN BUILDING BLOCKS

Proteins are macromolecules made up of lengthy chains of building blocks called **amino acids**. Actually, 20 different amino acids are used in proteins, all of which have a general structure as shown in Figure 2-1a. This common structure is shared by all the amino acids, but each has a specific **side chain** (labeled R) that makes it different (Fig. 2-1b).

The side chains of amino acids are extremely varied. Some are very acidic (they tend to lose a hydrogen ion [H^+]). Some are basic (they attract an extra H^+). Some dissolve easily in water (hydrophilic), and others act like oil in water (hydrophobic—water hating). Some amino acids have ring structures (**aromatic**). Two amino acids have side chains containing sulfur, which gives decaying meat its awful odor. Structures of all the amino acids are given in the Appendix.

PROTEIN STRUCTURE

Primary

Proteins are made of long strings (long chains) of amino acids, linked together with a very strong covalent bond (see Chapter 1), the **peptide bond** (Fig. 2-2). Before this chain of amino acids folds up to give it structure, it is called a **polypeptide chain**. The many intricate structures of proteins are specified by the order in which the amino acids are placed in the polypeptide chain. This sequence of amino acids in a polypeptide chain is referred to as the protein's **primary structure** (Fig. 2-3).

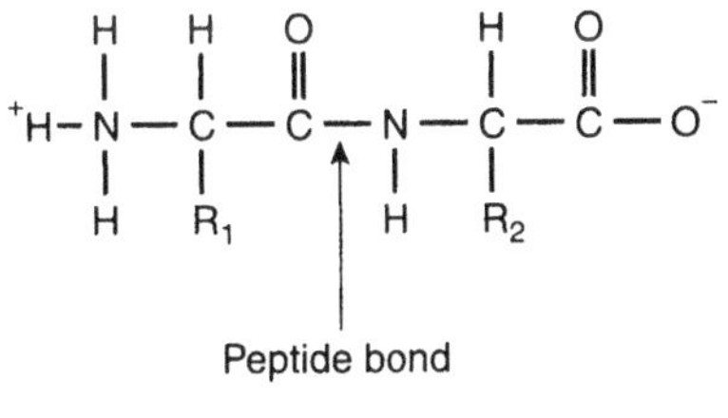

Peptide bond

Figure 2-2 Structure of a peptide bond within a polypeptide chain (*arrow*). This is the link between adjacent amino acids. The bond forms between the carbon of one amino acid and the nitrogen of the next, forming a chain of amino acids (a polypeptide chain; see Fig. 2-3). This is a covalent bond and is very strong.

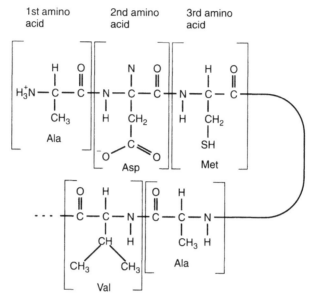

Figure 2-3 An example of a portion of a polypeptide chain. In a protein, the sequence of amino acids linked together by peptide bonds is called the primary structure of the protein. (Amino acids shown here: Ala = alanine; Asp = aspartic acid; Met = methionine; Val = valine.

Secondary

After the long chain of amino acids has been made, each chain is folded in a particular fashion according to the sequence of amino acids. This is called the **secondary structure** (Fig. 2-4b). This folding of the polypeptide chain is caused by interactions between the amino acids, generally hydrogen bonds (H-bonds), and by hydrophobic interactions of the polypeptide chain with the environment surrounding it. Sometimes the initial folding of the polypeptide chain is in the form of a spiral-like structure called a **helix**. The folded polypeptide chain is called an **alpha (α) helix** and is very common in many proteins (Fig. 2-5).

The helical structure predominates in hair proteins and some other structural proteins, and it is present to some degree in most proteins. In hair, helical strands have a tendency to wind around each other, much like the strands in a rope, to give lots of strength to the whole structure but retain great flexibility. Sometimes, disulfide (S-S) bonds form between the hair strands (fibers), which cause them to curl. When we get a hair perma-

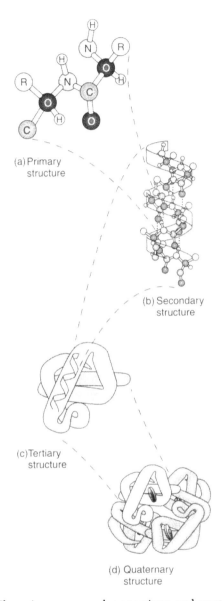

(a) Primary
structure

(b) Secondary
structure

(c) Tertiary
structure

(d) Quaternary
structure

Figure 2-4 The primary, secondary, tertiary, and quaternary structure of proteins. (**a**) The primary structure is the sequence of amino acids linked by peptide bonds. (**b**) The secondary structure is the folding of the amino acids. (**c**) The tertiary structure is the coiling of the coils, and (**d**) the quaternary structure occurs when more than one coiled coil is bound together. See text for details.

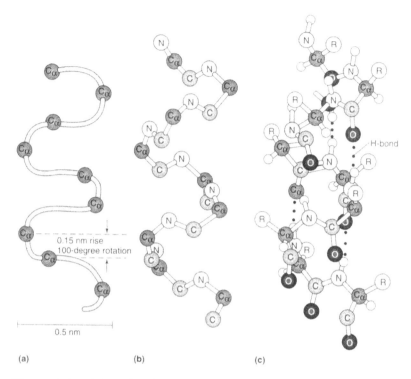

Figure 2-5 Alpha (α)-helical structure found in proteins. Drawing (**a**) shows only the α-carbon atom $C_{\cdot_{\alpha}}$ of each amino acid in a helical form. Drawing (**b**) shows all the atoms in a helical backbone (N = nitrogen). Drawing (**c**) shows the additional atoms as well as the side chains (**R**) in the amino acids. Note the hydrogen-bonds (H-bonds), which hold the amino acids in the α-helical structure.

nent or use a curling iron, we are merely breaking and remaking some of the S-S bonds (Fig. 2-6).

Another very common protein structure occurs when the amino acid chains run parallel with each other for a short distance. These sheet-like structures (beta [β] sheets) often contain many amino acids with hydrophobic side chains (Fig. 2-7). At times, these β-sheet regions form cavities and areas in proteins that are very hydrophobic and attract other hydrophobic molecules. These regions are also useful when proteins mix with the membranes of cells, which are made up of fatty molecules.

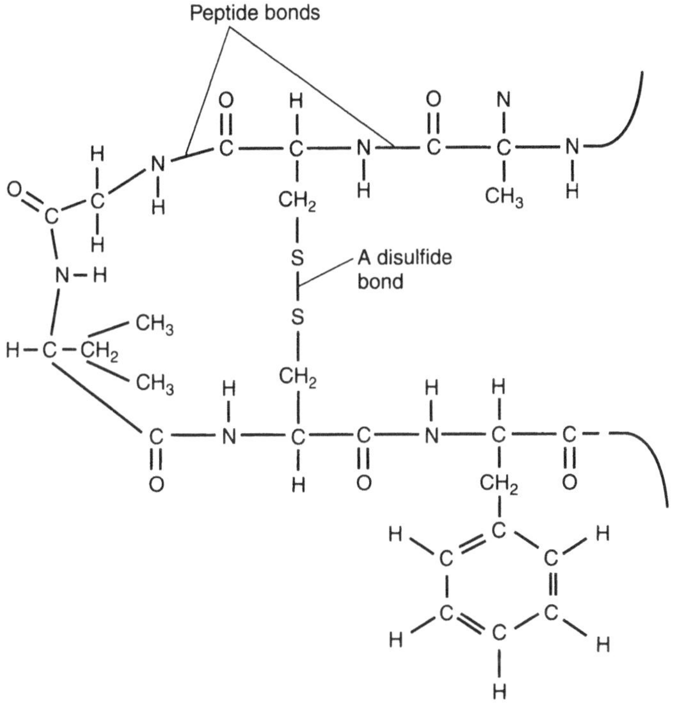

Figure 2-6 Diagram showing several amino acids linked together with peptide bonds, with two of the side chains forming a disulfide (S-S) bond, which gives the structure additional strength.

Tertiary

When the folded regions of the amino acid chain fold on themselves, this is called the **tertiary structure** (Fig. 2-8). At this structural level, the helices and sheets are folded and kinked in specific ways, allowing various portions of the amino acid chain to come in contact with one another. This refolding of the helical and sheet-like structures often causes the proteins to become very compact and tight, which often helps them to keep out water. All the reasons for the folding of proteins are not yet understood, but it is clear that the correct folding is critical to a particular protein's function.

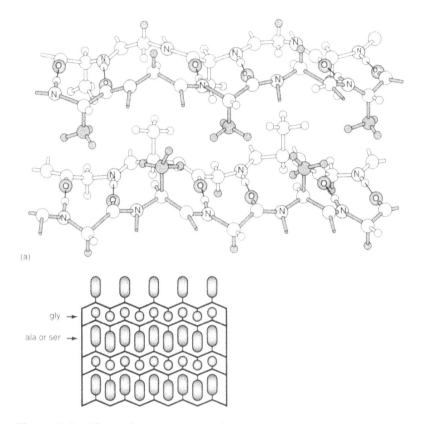

Figure 2-7 The β-sheet structure of proteins. Drawing (**a**) is a "ball and stick" structure showing how the polypeptide strands have a sheet-like structure relative to each other. Drawing (**b**) shows a more detailed view of the way the side chains of the amino acids interdigitate when the sheets are layered. (ala = alanine; gly = glycine; ser = serine.)

WHAT PROTEINS DO

Proteins do lots of different things in a cell. The function of each protein depends on its structure, which in turn depends on the way in which the folding of the polypeptide chain occurs. And the folding of the polypeptide structures of all proteins depend on the sequence of amino acids in the polypeptide chain. So the sequence of amino acids critically determines the function of the protein.

Tobacco mosiac coat protein

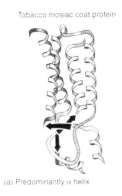

(a) Predominantly α helix

Immunoglobulin

(b) Predominantly β sheet

Pyruvate kinase

(c) Mixed α helix and β sheet

Figure 2-8 Different kinds of tertiary structures of proteins. Drawing (**a**) shows a protein with predominantly α-helical structure. Drawing (**b**) shows a portion of a protein containing mainly β-sheet structure. Drawing (**c**) is an example of a portion of a protein that has a mixture of both the α-helical and β-sheet structures. Some proteins have portions that are more α-helical, and other portions have more β sheet.

Enzymes

By far the most important function of proteins is to serve as enzymes. An enzyme is a very precisely made protein that causes chemical reactions in the organism to happen much faster than they would normally. In chemical terms, these are called **catalysts**. The reason we need enzymes is to speed up (**catalyze**) the chemical reactions that take place. Enzymes can make the chemical reactions proceed thousands and even millions of times faster than they normally would at body temperatures. For instance, when we eat food, enzymes cause the food to be broken down into substances that can be used by our body.

Enzymes are specialists. The activity of an enzyme is a direct result of the precisely made protein structure that allows it to function very efficiently, but only for a very specific reaction or group of reactions. Enzymes are capable of recognizing and, when necessary, rejecting almost-identical structures that are different only in the placement of a single carbon (C) or a hydrogen (H) atom. This amazing specificity has often been pictured as a lock-key fit (Fig. 2-9).

If the order of the amino acids in the polypeptide chains is not exact, the specificity or activity of a particular enzyme may be lost, which may kill the cell. If we multiply this required accuracy by the thousands of proteins that perform the multitude of tasks essential for life, we can get a glimmer of how critical it is that the genetic message, which determines the order of the amino acids, be kept exact.

Structural Proteins

Proteins are also used for structural purposes. In our bodies, muscles, tendons, and ligaments literally hold our frame together. In addition, our skin, hair, and fingernails all are primarily protein. Each protein is masterfully designed and exactly made to provide the necessary function. For instance, ligaments are made primarily of **collagen**, which is a triple-stranded helical fiber. It takes many of these triple-stranded fibers wound around each other to make a tendon or a ligament (Fig. 2-10).

The tensile strength (the amount of pulling necessary to break it) of collagen is enormous; still, it can be broken, a fact to which many football players and skiers can attest. Collagen has tensile strength per unit mass far greater than steel; yet, it is

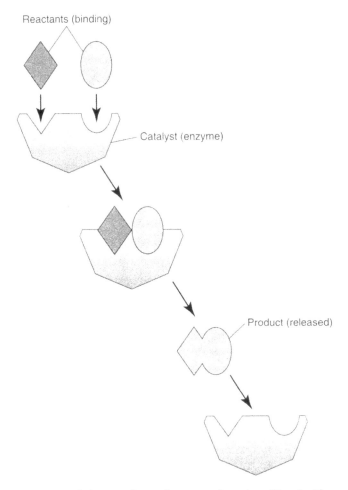

Figure 2-9 Lock-key analogy of enzyme function. That is, if two molecules are to be bound together, the enzyme helps this reaction by fitting one reactant molecule into one site on the enzyme and fitting the other into an adjacent site on the enzyme. By bringing the two reactant molecules together, positioned in the right way, the enzyme greatly improves the rate of product formation. The product is made of the two reactant molecules coupled together.

much more flexible. It even has portions of very flexible protein to provide additional stretching and bending power. Collagen is but one example among many of the diverse ways in which proteins can be designed and made for specialized purposes.

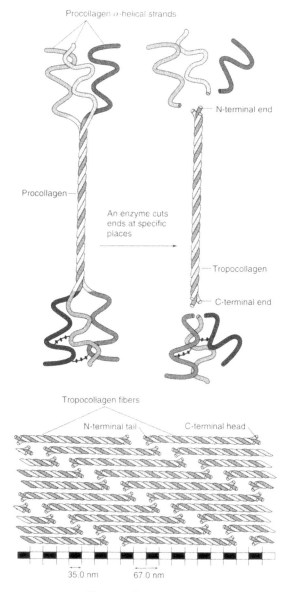

Figure 2-10   Diagram of how collagen, a structural protein, is made. Three procollagen α-helical strands are wound together to give procollagen. The unnecessary ends of the procollagen are then cut off by a specific enzyme, leaving the tropocollagen strand. Many of these are then overlapped together in a long fiber to form collagen, a portion of which is diagrammed here.

Transport Proteins

Transport of nutrients and oxygen is another major duty of proteins. One of the proteins studied in most detail in this arena is **hemoglobin**, the carrier of oxygen (Fig. 2-11). This amazing molecule is made of four polypeptide strands and is able to

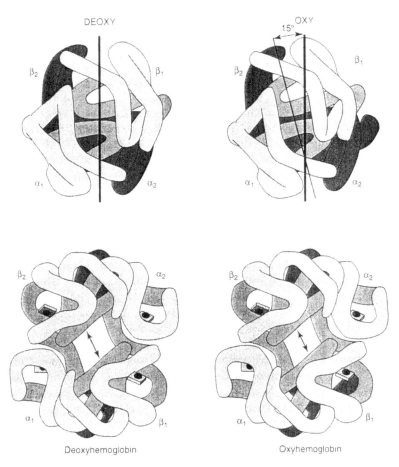

Figure 2-11 Hemoglobin, the carrier of oxygen. Hemoglobin is made of four separate protein chains, two α chains and two β chains. In the deoxy (without oxygen) form, the chains are positioned quite symmetrically with a large hole between them in the middle. When oxygen is added (the little platforms and balls shown), the subunits tilt about 15 degrees relative to each other, and the hole in the middle becomes smaller. These slight changes in structure make it possible for hemoglobin to get oxygen from the lungs and take it to the cells and release it.

carry four molecules of oxygen. Hemoglobin not only must pick up the oxygen in the lungs, but it must drop it off in the capillaries and the muscles where it is needed, and then transport carbon dioxide (CO_2) back to the lungs to be expelled. The finely tuned mechanism that allows this to happen has been studied over the last 60 years and is now understood in great detail. Needless to say, the precise structure of hemoglobin is critical. Any alteration in its structure can have fatal consequences, as we shall see in Chapter 4.

Signal Proteins

Another major protein function is messenger service. An example of this is with insulin, which helps regulate blood sugar amounts. It does this by telling muscle and fat cells that there is a supply of food (sugar) present and helps them to absorb this sugar. Many of our hormones are proteins and specifically work to keep our body functioning by providing signals to various cells that enable our bodies to regulate temperature, activity, stress, pain, and other conditions.

Glycoproteins and Lipoproteins

Finally, we should mention that proteins are often coupled with sugars and fats to form complexes that have even more variability. Proteins coupled with sugars are called **glycoproteins**, and these are amazingly diverse (Fig. 2-12). One of their primary functions is to serve as cell markers so that various kinds of cells can be recognized. This is especially important to our immune system, which has to discriminate between "self" and "nonself" (foreign) cells. In addition, glycoproteins perform intricate functions with membranes and signal pathways, which could be done in no other way.

Lipoproteins are made of lipids (fats) and proteins and are primarily designed to transport fats in aqueous (water) systems, such as in our bloodstream. These fats are carried from our digestive system to cells to be used for food and for energy storage. The protein component serves to provide a "shell" to interact with water and in which to hide the fats.

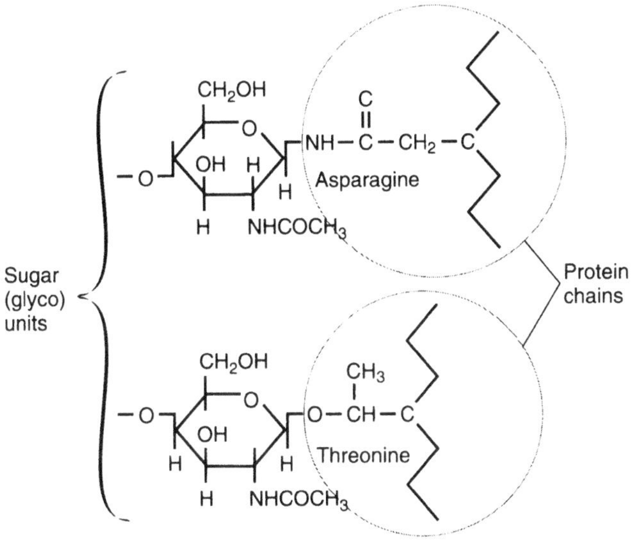

Figure 2-12   Examples of two glycoproteins that are formed by sugar (glyco) units being attached to protein chains through amino acid side chains. These structures are especially useful in cell wall and cell membranes.

SUMMARY

Proteins come in many varieties, sizes, and shapes and are made of 20 amino acids. All amino acids have similar structures, but have different side chains, which give them specific properties. The amino acids are linked together in the order dictated by the genetic information, into long chains called polypeptide chains. These polypeptide chains are folded into helical or sheet-like structures. Then these structures are often folded again to give the protein the exact structure it needs for its particular function.

Proteins function as enzymes (catalysts), structural elements, and signal devices. They can also be combined with sugars to make glycoproteins (often used as markers for cells) and with fats to make lipoproteins (often used as transport device for fats).

3

NUCLEIC ACIDS

WHAT YOU WILL LEARN IN THIS CHAPTER

- How deoxyribonucleic acid (DNA) was found, its structure, and why it is important
- How DNA makes copies of itself
- How ribonucleic (RNA) is made
- What roles DNA and RNA perform

In the movie, *Jurassic Park*, the story line develops around fragments of DNA from prehistoric dinosaurs that were cloned and ultimately develop into fully functional creatures of grand proportions. This is all supposed to result from using just *some* of the DNA from these extinct creatures. Could this really happen? Is it possible to use DNA from a living creature and get a new, identical living creature? What are the limitations of this approach? To understand this, we need to dig a bit into the storehouse of genetic information, DNA, and find out how it tells cells what do.

All the genetic information for cells is contained entirely in the DNA. DNA, as noted in Chapter 1, is made of a long string of four deoxyribonucleotides, abbreviated by the letters A, G, C, and T (see Figure 1-11 for the complete structures of the nucleotides). All the information necessary to make complete cells and bodies is contained in these long strings of nucleotides. So it seems as if it may be possible to make a new dinosaur if we had all of its DNA. But we will discuss this in more detail in Chapter 10.

A BRIEF HISTORY OF DNA

DNA has not always been known to be genetic material. Actually, in 1869, Johann Friedrich Miescher discovered DNA and named it **nuclein**, because it was isolated from the nucleus (central core) of cells. However, its function was completely unknown and remained so for almost a century thereafter.

It is interesting that only 3 years previous to Miescher's discovery, Gregor Mendel, an Austrian monk, was just completing his study of pea genetics and had deduced the fundamental laws of genetics (the scientific study of heredity). Mendel had carefully studied seven characteristics of pea plants (Fig. 3-1). To do so, he crossed plants having a certain form of one characteristic (e.g., a green seed) with others having a similar, but different characteristic (e.g., a yellow seed). First, Mendel waited until the offspring plants had self-fertilized. Then he counted the number of offspring that displayed each kind of a particular characteristic. He discovered that the ratios of the forms of the characteristics occurring in the offspring were the same for each of the seven characteristics—approximately 3:1.

To explain his results, Mendel postulated that two factors control the characteristics and that one factor dominates the other. This laid the foundation for the theory of inheritance, but it was not well received. Mendel's idea that sex cells (eggs and sperm) are able to transmit these heritable characteristics was controversial. He shared his ideas with other scientists, but they didn't have much faith in them at that time. Not until the early 1900s did scientists revisit the inheritance problems in plants and confirm Mendel's work, showing that inheritance factors were related to the portion of the cells called **chromosomes**. Chromosomes are mixtures of nucleic acids and proteins found in cell nuclei.

Even with this information about inheritance, however, Mendel's insightful discoveries and Miescher's discovery of nuclein did not come together. Although the chromosome was known to contain nuclein—or nucleic acid, as it came to be known—early scientists thought that nucleic acids were not complex enough to carry a lot of information. To early investigators, it seemed impossible that such a biological macromolecule containing just four different building blocks would be able to contain much useful information. They believed it would be like trying to write a book using four letters of the alphabet.

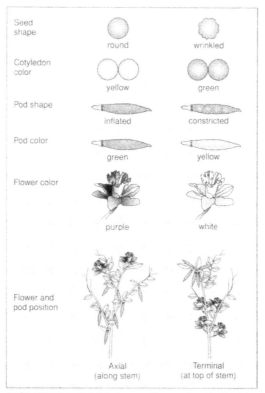

(a) Six of the characteristics of pea seedlings analyzed by Mendel

Parental characteristics	F_1	Number of F_2 progeny	F_2 ratio
Purple x white (flowers)	Purple	705 purple, 224 white	3.15 : 1
Yellow x green (cotyledons)	Yellow	6022 yellow, 2001 green	3.01 : 1
Round x wrinkled (seeds)	Round	5474 round, 1850 wrinkled	2.96 : 1
Inflated x constricted (pods)	Inflated	882 inflated, 299 constricted	2.95 : 1
Long x short (stems)	Long	787 long, 277 short	2.84 : 1

(b) Results

Figure 3-1   Mendel's experiment. (F_1 = first generation of offspring; F_2 = second generation.)

It was much easier, then, to suppose that proteins were carrying the complex genetic information. After all, proteins contain 20 different amino acid building blocks, not just 4. And proteins were found in the nucleus of cells as well. So it seemed logical that proteins had much more capacity to carry the genetic information, since 20 "letters" give many more possibilities than 4 would. Still, this turned out to be a wrong conclusion.

In 1945 near the end of World War II, Oswald Avery, Colin MacLeod, and Maclyn McCarty, three scientists at the Rockefeller Institute in New York, showed after a decade of careful experimentation that, by transferring only DNA, genetic information can be taken from one strain of bacteria and given to another (Fig. 3-2). This was the first solid evidence that DNA, having only four nucleotides, actually contains genetic information.

The results of Avery, MacLeod, and McCarty's experiments showing that DNA was the genetic material were startling, but the message didn't sink in to the scientific community until much later. Other experiments were performed and other approaches were taken, each adding more proof.

Early researchers didn't pay much attention to the structure of DNA, because it had been considered uninteresting. Most thought that all four of the nucleotides were placed in DNA in a constant, repetitive order and in equal amounts. However, in 1950, Erwin Chargaff discovered that the amount of G equaled that of C, and the amount of A equaled that of T (see next paragraph and Table 3-1). This was a critical piece of information, which set the stage for the discovery of the structure of DNA and allowed us to understand how the genetic message it carried was transferred from DNA to DNA as cells reproduced themselves.

Table 3.1 Ratios of Adenine, Thymine, Guanine, and Cytosine Obtained by Chargaff

SPECIES	A	T	G	C
Homo sapiens	31.0	31.5	19.1	18.4
Drosophila melanogaster	27.3	27.6	22.5	22.5
Zea mays	25.6	25.3	24.5	24.6
Neurospora crassa	23.0	23.3	27.1	26.6
Escherichia coli	24.6	24.3	25.5	25.6
Bacillus subtilis	28.4	29.0	21.0	21.6

Note that the percentages of adenine (A) and thymine (T) in each species are similar, as are the percentages of cytosine (C) and guanine (G).

DNA STRUCTURE

As noted in Chapter 1, DNA is made of nucleotides, which are sugar molecules with nitrogen-containing bases attached to

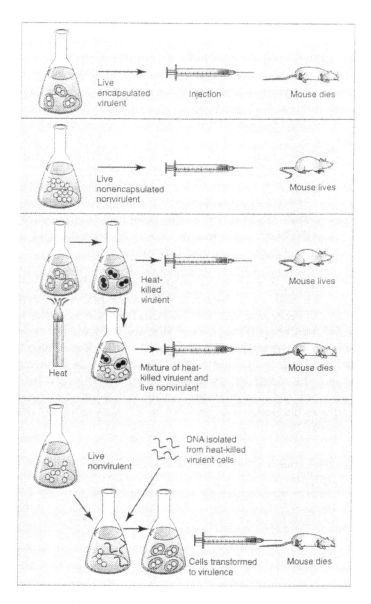

Figure 3-2   The Avery, MacLeod, and McCarty experiment. Since the live nonvirulent cells would not kill a mouse, what factor was transferred from the heat-killed virulent cells that made the live nonvirulent cells kill mice? By isolating various fractions of the heat-killed cells, these scientists ultimately isolated the DNA fraction as the fraction that was responsible for the transformation of the nonvirulent cells into virulent cells. This was the first time that DNA was clearly shown to have the ability to genetically alter cells, suggesting that DNA contained genetic information.

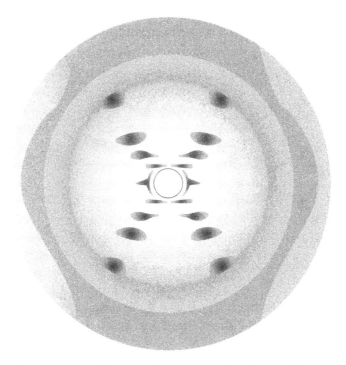

Figure 3-3   This is a diagram of x-ray scattering from a fiber of DNA. Watson and Crick used such a fiber pattern to determine the structure of DNA. The helical structure is shown by the X-shaped pattern of the spots on the film. The distances between the spots tell scientists how much distance there is between the nucleotides in the helical strands.

them. Of the four bases used in DNA, two are single-ring struc- tures and two are double-ringed. The single-ring structures are **cytosine** (C) and **thymine** (T). The double-ring structures are **adenine** (A) and **guanine** (G). These structures are shown in Figure 1-12. When the sugar, the phosphate group, and a base are all bound together, we have deoxyribonucleotides.

Using some x-ray patterns of fibrous DNA as a basis (see Fig. 3-3), James Watson and Francis Crick made a detailed model of the structure of DNA, which showed it to be a double- stranded helix (Fig. 3-5).

One of the key features of the model was that A-T and G-C bind together, making one strand of the DNA *complementary* to the other. Thus, if one strand contains the sequence AGCT, the bases on the other strand would be TCGA. The bases on the adjacent strands are attached by means of hydrogen bonds (H-bonds). Figure 3-4 shows the structure of the paired bases.

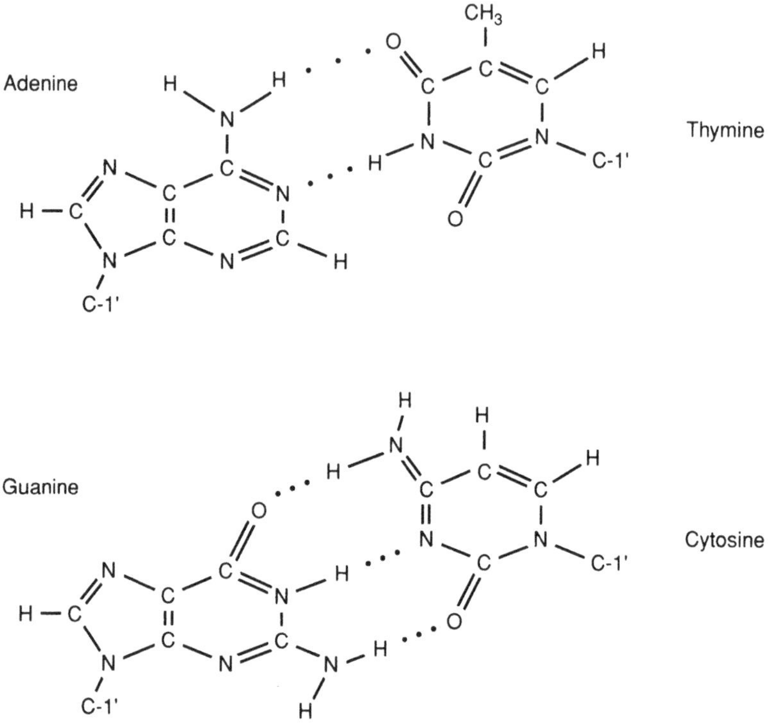

Figure 3-4 Structures showing base pairing (hydrogen-bonding) of adenine (A) and thymine (T), and guanine (G) and cytosine (C). These base pairs are very close to identical distances across, making them ideal for attaching the two strands of DNA together.

The x-ray patterns suggested that the DNA strands were coiled in a helical (spiral) fashion (Fig. 3-5). The discovery of this double spiral, called a **double helix**, earned Watson and Crick a Nobel Prize and really kicked off the detailed study of genes. The double-strandedness of DNA, together with the ability of each strand to be complementary to the other, was the key to understanding how genetic information might be stored and reproduced.

How are the nucleotides linked together in a DNA molecule? The covalent bonding of the strand is not between the nitrogen-containing rings of the bases, but between the sugar molecules through the phosphate groups. This happens between the fifth

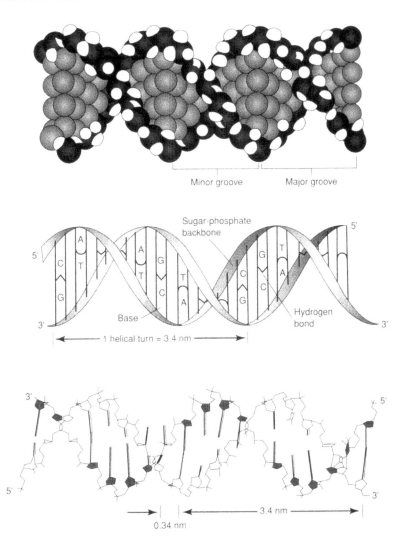

Figure 3-5   Three representations of the DNA double helix. Note the presence of the major groove and minor groove in the structure. Water often binds in the minor groove. Also, note that the base pairs are almost perpendicular to the axis of the helix (lower drawing). The sugar-phosphate backbone is on the outside, and the bases are located in the interior of the double helix.

carbon (5′ carbon) of the deoxyribose sugar on one molecule and the third carbon (3′ carbon) of the next sugar (Fig. 3-6). Thus, each deoxyribose molecule then has two phosphate groups attached to it, one at the 3′ and one at the 5′ carbon. The

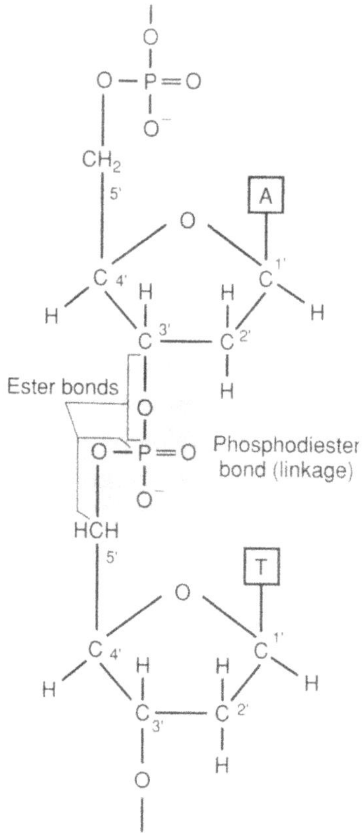

Figure 3-6 The sugar-phosphate (phosphodiester) bonds that link two deoxyribose or ribose sugar units together. These bonds link large numbers of the sugars together in a sugar-phosphate backbone making a DNA (or RNA) strand.

result is that the phosphate group becomes the link between the deoxyribonucleotides. This covalent bond between the carbons on the sugars and the phosphate group is called a **phosphodiester bond** or linkage (see Fig. 3-6).

REPLICATION

When cells are dividing, new DNA must be made. This is done by adding a single nucleotide to the end of a growing chain. This addition always occurs at the 3′ end of the chain. So the new

chain always starts out with the 5' nucleotide, and then additional nucleotides are added until it is finished. When we talk
about the sequence of a chain of nucleotides, we always write it
from the 5' end. The process of DNA making new DNA strands
is called **replication** (Fig. 3-7).

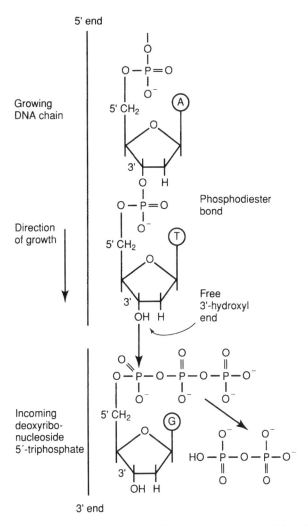

Figure 3-7   The birth of a strand of DNA. New nucleotide triphosphates are added to the 3' end of the growing chain. Growth is always
in the 5' to the 3' direction. When the new nucleotide is added, the two
extra phosphate groups are broken off, which gives the energy needed
to make the new link between the new nucleotide and the 3' end of the
growing strand.

We now can see how a long chain of nucleotides might be put together in a random sequence. But how is a certain sequence specified? The genetic message is found in the sequence of nucleotides initially present in the DNA. This sequence must be specified in each of the new pieces of DNA with exactness. So there must be a mechanism by which the sequence of nucleotides in new DNA can be made to match that of its parents with exacting precision. Here, the beauty of the double-stranded DNA molecule comes into play.

Because the two strands of DNA are complementary to each other, one strand becomes a pattern (**template**) for the other strand. Then, when one strand has an A, the other strand would have a T. Likewise, where a G is present on one strand, then a C would occur on the other strand. Thus, if one strand contains a series of bases 5′ AGGCTTACC, the other must have 3′ TCC-GAATGG as its sequence. If we think in terms of the first sequence being a master coding strand, then it would specify the sequence of the complementary strand (Fig. 3-8). Therefore, one strand of DNA can be used to make the other strand, which would have a matching, yet complementary, sequence (see Fig. 3-8).

To make new DNA for a daughter cell, the parent DNA strands are separated. Each of them is used as a template for a complementary strand, resulting in two identical molecules of double-stranded DNA. This replication process is carried out by **DNA polymerases,** which are enzymes specially designed to read the sequence of bases on the template strand and place a complementary base in that order on the new strand.

Replication takes place very rapidly, and sometimes a wrong nucleotide is inserted. So other proteins move along the DNA "proofreading" the new strand for possible mismatches and fixing them. In this way, the genetic message is kept practically free from error—only making an error once for every 10 billion nucleotides linked together! In other words, if this book were written in DNA code, only one typographical error would appear in one letter of one word out of 5000 different books this size.

Another important feature of the double-helical nature of DNA is that the structure is "locked" in. This feature makes the sequence contained in the DNA molecule very stable and prevents accidental changes. If one strand is somehow damaged, the other strand can be used to repair it, keeping the message the same. This makes DNA an ideal molecule in which to **archive** (store) genetic information.

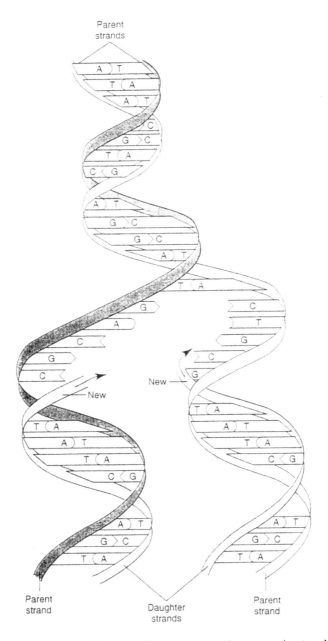

Figure 3-8 DNA replication. Two new strands are made simultaneously by special enzymes called DNA polymerases. The DNA is unwound, and each of the parent strands is used as a template to make an identical copy of the complementary strand. The result is two double strands of DNA, rather than just one. This process takes place every time a new cell is made.

One final feature of DNA needs to be emphasized. Although it is easy to show two strands of DNA parallel on paper, the natural form is helical. The twisting of the strands of nucleotides causes DNA to look very much like a spiral staircase, with the steps being represented by the base-paired bases and the bannister region by the sugar-phosphate backbone. Sometimes, DNA takes other forms, making loops and bulges, but generally the Watson-Crick structure is found.

The helical structure also makes DNA convenient for **packaging** in the nucleus. The helical strands can be folded and twisted back on each other to make it very compact. DNA is a very long molecule, so it has to be folded and refolded with great care to fit within the cells. The helical structure is ideally suited for this purpose, because it is fairly flexible, yet compact. The packaging in more complex cells is often done by wrapping the DNA around proteins, which allows the DNA to fold into tightly folded structures of protein and DNA called **nucleosomes** (Fig. 3-9). Clearly, good packaging is essential, especially in a bacterium or virus, which has a lot of DNA.

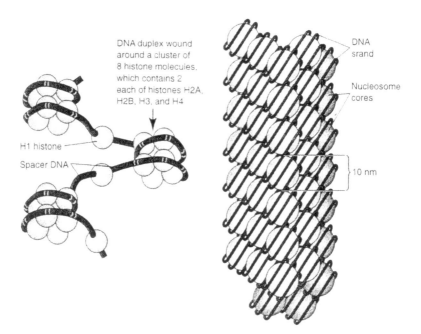

DNA duplex wound
around a cluster of
8 histone molecules,
which contains 2
each of histones H2A,
H2B, H3, and H4

DNA
srand

Nucleosome
cores

H1 histone

Spacer DNA

10 nm

Figure 3-9   Structure of nucleosomes.

DNA's structure is ideally suited for its purpose. The long series of nucleotides in a row allows information to be coded by merely changing the sequences of the letters, much like a four-letter alphabet. Because of the double-stranded nature of its structure, genetic information can be stored indefinitely. It can be reproduced with ease, repeating the genetic message that it obtained from its ancestral DNA. In turn, it can replicate itself by transferring this same information to its progeny (offspring) with great accuracy. However, if it is changed or modified before it is replicated, then the modified information is expressed and substantial changes can be made in the proteins produced. We will discuss this more fully in Chapter 4 after we learn more about how the genetic information is used.

RNA

Ribonucleic acid (RNA) is a bit different from its cousin, DNA. It is made out of the same kind of bases as DNA, except that in place of thymine (T), RNA uses uracil (U) (Fig. 3-10). In addition, the oxygen that was missing from the second (2') carbon of the ribose sugar in DNA is present in RNA. Therefore, the ribonucleotides have an O-H (oxygen bound to the carbon) group at 2' carbon on the sugar, whereas the deoxyribonucleosides have a mere hydrogen atom there (see Fig. 1-12). The oxygen atom is a lot larger than the hydrogen atom, so the diagrams we make don't always tell the whole story.

RNA differs from DNA in its structure and function. First, RNA is almost always single-stranded. Although two strands of RNA can join and form a double-stranded structure, this seldom happens naturally. RNA would rather stay single-stranded and fold this single strand back on itself in a variety of ways, giving double-stranded regions, loops, bulges, and other interesting structures. This helps protect the RNA from enzymes that chew it up. It also helps give some portions of the RNA a specific function. Second, RNA is much less stable than DNA. Perhaps this is because it is single-stranded—or at least has regions that are. This makes RNA more vulnerable to attack by **ribonucleases**, enzymes that specifically chew it up. In addition, the extra -OH group on the 2' carbon causes added chemical activity.

If the RNA doesn't have two strands, how is it made? It is made by a process called **transcription**, in which one strand of DNA is used as a template for the new RNA. Transcription

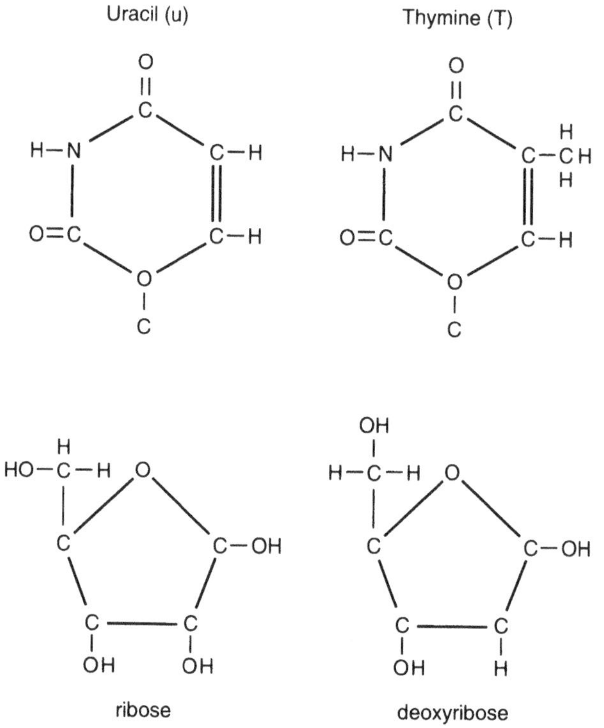

Uracil (u)   Thymine (T)

ribose   deoxyribose

Figure 3-10   Differences in the structure of uracil (U), which occurs only in RNA, and thymine (T), which occurs only in DNA are shown. Note also the difference between ribose (only in RNA) and deoxyribose (only in DNA), which lacks an oxygen at position 2′.

occurs in a way very similar to the way DNA replicates itself. The double helix of the DNA is opened up at a specific site, and one of the strands is copied in a complementary manner by a special enzyme, **RNA polymerase**. In this way, the genetic message, which is contained in the DNA, is transferred to the RNA (Fig. 3–11).

Transcription differs from replication in two ways. First, transcription of RNA generally takes place on only one of the DNA strands (the template strand), whereas replication takes place on both strands at the same time. This is understandable because it is difficult to see how both DNA strands would be useful in making a message (although this is done in some rare cases).

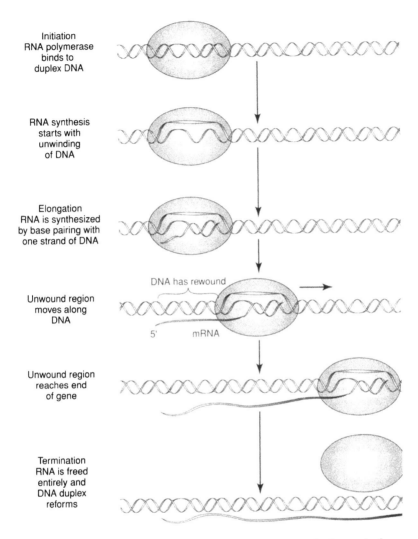

Initiation
RNA polymerase
binds to
duplex DNA

RNA synthesis
starts with
unwinding
of DNA

Elongation
RNA is synthesized
by base pairing with
one strand of DNA

DNA has rewound

Unwound region
moves along
DNA

5'   mRNA

Unwound region
reaches end
of gene

Termination
RNA is freed
entirely and
DNA duplex
reforms

Figure 3-11 RNA transcription. In this process, RNA is made from one of the DNA strands (the template) and is carried out by an enzyme called RNA polymerase. In the process, the DNA is unwound, and then wound up again after the new RNA is made. The diagram shows mRNA (messenger RNA) being made, the RNA that carries the genetic message to be made into proteins.

Second, transcription also takes place much more slowly than replication and does not contain a proofreading step. It is much less important for the transcribed RNA to be exact than it is for the replicated DNA to be exact. This is because the RNA that serves as the genetic message will be used only a few times and then discarded. So, the effort needed to proofread is not worth it. But transcription is tightly **regulated**, which means that the amount and kind of RNA that are made are carefully controlled. This regulation is one way in which the organism determines how many proteins of a certain kind are made.

There are three major types of RNA in a cell: messenger RNA, transfer RNA, and ribosomal RNA.

Messenger RNA

Messenger RNA (mRNA) is an exact copy of the genetic message on the DNA. It is transcribed from the template strand of DNA. The template strand is called the **negative (−) strand**, because the mRNA is always the **positive (+) strand**. After the mRNA has been transcribed, it is attached to the ribosomes (see explanation under Ribosomal RNA and in Chapter 4), where proteins are made. The mRNA carries genetic information out of the cell's nucleus (if it has one), where it is used by the ribosomes as the message that is translated into protein. In the case of higher-order cells (**eukaryotes**; larger, more highly organized cells, with nuclei and other organelles), the RNA is sometimes peppered with segments (**introns**) that are not to be translated. These introns have to be removed before translation can occur. In addition, mRNA molecules from eukaryotes always have a string of "A"s at the 3′ end, which is called a "poly A tail." Sometimes, especially in bacteria, the mRNA can code for more than one protein.

mRNA generally has a short lifetime. Once the message is translated on the ribosomes to which it attaches (discussed in Chapter 4), ribonucleases chop it into single ribonucleotides, allowing these units to be recycled back into other RNA molecules. This is one way in which the number of proteins in cells is controlled so that too many copies of the same protein are not made.

We can see that mRNA is not just a long string of nucleotides. It folds back on itself in some very complicated structures. We are only just beginning to understand the purpose of this folding (Fig. 3-12).

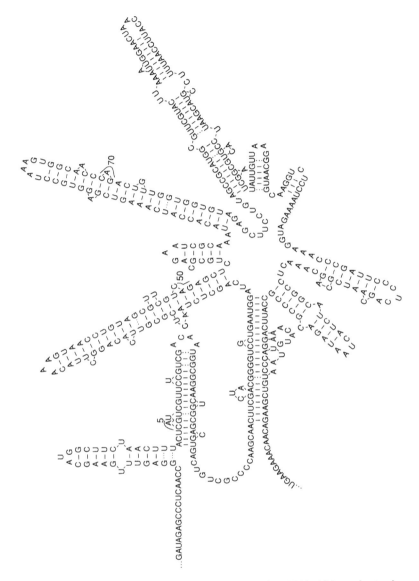

Figure 3-12 Structure of a messenger RNA (mRNA). Although single-stranded, RNA folds back on itself in a series of hairpin-like loops. Note that there are some bases that "bulge" out and some looped sections within double-stranded regions. RNA structure can be very complicated, even though it is made by a single strand.

Transfer RNA

Transfer RNA (tRNA) is a very small RNA molecule, generally containing 75 to 85 nucleotides. The tRNA polynucleotide strand is tightly folded and twisted to give it a structure that looks a little like a fat, upside-down L, as shown in Figure 3-13. tRNA is used as a shuttle service for amino acids. There is at least one different tRNA for each of the 20 amino acids. A special enzyme selects a tRNA for a specific amino acid and then finds that amino acid and attaches the amino acid to the tRNA. The tRNA then carries that amino acid to the ribosome and places it in the proper position in a protein chain that is being made. These tiny RNA molecules are very abundant and have

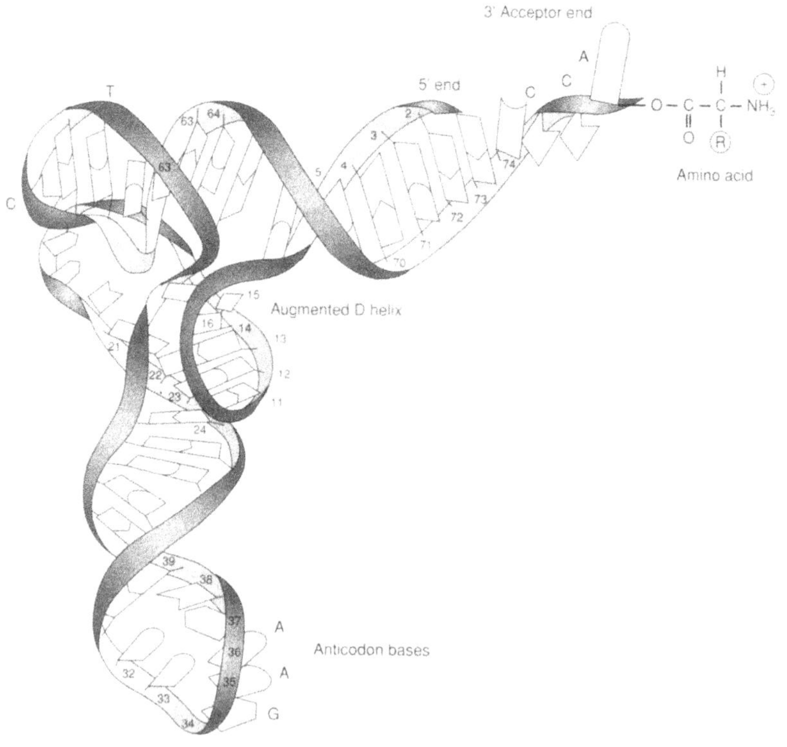

Figure 3-13 Schematic diagram of a transfer RNA (tRNA) with an amino acid attached. The ribbon represents the phosphodiester backbone. All tRNA molecules have very similar structures, but with subtle differences to allow them to receive only a particular amino acid.

subtle differences in structure that allow them to be recognized by the enzyme that attaches them to their specific amino acid.

In addition to its being a shuttle service, tRNA must have two unique characteristics.

- It must have particular features of structure that allow an enzyme to attach only a specific single amino acid to it and no others.
- It must be able to identify the unique code for that amino acid on the mRNA.

Moreover, the entire tRNA-amino acid complex must be able to attach to a specific region on the ribosome. So, tRNA, even though small, has several critical functions to perform.

Ribosomal RNA

Ribosomal RNA (rRNA) is by far the most abundant RNA in cells. It is found within the massive structures called **ribosomes**. Ribosomes are made of one large and one smaller subunit and are the factories in the cell in which new proteins are made. Ribosomes are composed of rRNA strands and many different proteins. Each ribosomal subunit has a long piece of rRNA. The RNA of the larger subunit is almost twice as large as that of the smaller subunit. In addition, the large subunit contains a very small piece of rRNA (Fig. 3-14).

rRNA contains lots of double-stranded regions and hairpin loops and bends. These are folded around the proteins in a particular way and provide binding sites for both the tRNA molecules and the mRNA. It is not yet known exactly where all these sites are on the ribosome, but it is clear that rRNA has much to do with the production of new proteins, as described in more detail in Chapter 4.

Some regions of the rRNA in bacterial ribosomes are identical with those found in ribosomes from all other cells, including those from humans. This suggests that these particular regions are especially important in the structure or function of the ribosomes. It is interesting to note that many antibiotics that kill bacteria and other cells do this by attaching to specific sites on the ribosome and disrupting its protein-making function.

We now see that RNA comes in various sizes, shapes, and packages and is used in cells for messenger and shuttle work and for helping in the complex process of making new protein. Although RNA is transcribed from the DNA and contains the

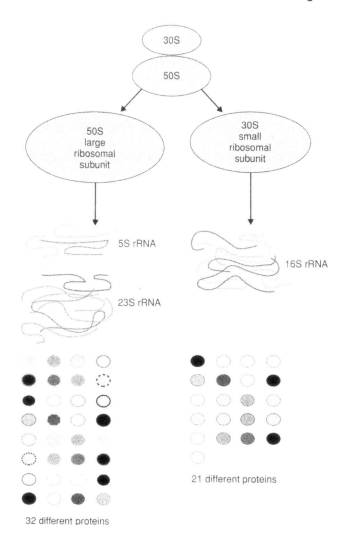

Figure 3-14 The bacterial ribosome. All living cells contain ribosomes—where proteins are made. Ribosomes are made up of ribosomal RNA (rRNA) and proteins. As shown schematically, there are two different ribosomal subunits—small and large. The small subunit (30S) has one strand of rRNA (16S; "S" refers to the size of these particles when determined in a centrifuge) and 21 different proteins, all bound together in a very specific way. Similarly, the large subunit (50S) contains two pieces of rRNA (23S and 5S) and 32 different proteins. Both subunits are essential and work together to make the ribosome function.

base sequence specified by DNA, it is not the archive (storage place) of this genetic information, but rather helps in the process of translating the genetic message into protein. Only in RNA viruses does RNA have archival duties.

Viral RNA

Many viruses contain RNA, not DNA. A number of these viruses cause influenza, common colds, some kinds of cancer, and AIDS. The RNA within these viruses not only has to provide the message to make the proteins that the virus needs, but also may have other structural or functional duties. Viral RNAs fall outside of the mRNA, tRNA, and rRNA categories and are generally just termed viral RNAs.

SUMMARY

The small building blocks of DNA and RNA, nucleotides, are used to make long strings of nucleic acids. DNA contains A, G, C, and T and is almost always found in a double-stranded, helical form. RNA contains A, G, C, and U and is almost always single-stranded. RNA has a great tendency to fold back on itself in hairpin loops and other complex structures. The process whereby DNA makes more DNA is called replication. Transcription is the process by which RNA is made from the DNA template (–) strand. Most RNA is found as messenger RNA (mRNA), transfer RNA (tRNA), and ribosomal RNA (rRNA). Viral RNA is found in some viruses, although most viruses contain DNA.

II
HOW LIVING THINGS ARE CHANGED

4

MAKING AND ALTERING PROTEINS

WHAT YOU WILL LEARN IN THIS CHAPTER

- How DNA makes RNA
- How ribosomes make protein from the genetic message contained by the RNA
- How the number of proteins is carefully regulated
- How the exact sequence of amino acids is needed to make functional proteins (changes do happen!)
- How harmful diseases cause many mutations

We have learned that proteins have various structures, which allow them to perform specialized duties within living systems (e.g., our bodies). We have hinted that unless each amino acid is placed at the precisely correct position in the protein, the protein is useless. So how are these amino acids put in the correct sequence? And what's this code we have been talking about? Let's find out.

THE CENTRAL DOGMA

Why do we have both DNA and RNA? Scientists in the 1950s asked this question in trying to unravel how genetic information was used. Francis Crick developed a statement that he called the Central Dogma: DNA transfers its genetic information to RNA, which in turn transfers that information to proteins.

So, proteins are the final receivers of the information contained in the genes and do not transmit information back to nucleic acids. The big question is how this information is transferred.

To make useful proteins, each amino acid building block has to be placed in a specified order in the chain of amino acids. Somehow this placement must be specified by the DNA. But how is this done? To understand this, we must revisit DNA and mRNA (messenger RNA) in a little more detail.

TRANSLATION

The Genetic Code

DNA is made of four nucleotides, A, G, C, and T. Just as with the letters of the alphabet, it is easy to see how these four letters can make a limited number of "words." However, there are 20 different amino acids in proteins and only four nucleotides in DNA. So information stored in words made of a four-letter alphabet must be used to specify the precise order of each of the amino acids in the protein strands. To do this, the DNA uses the **genetic code**.

The genetic code uses three letters of the DNA four-letter alphabet in various orders to represent each amino acid. When transcribed into mRNA, these three-letter sequences are called **codons**. Since four different letters in groups of three give more than 20 possible codons (actually 64), some amino acids are specified by more than one codon. Also, three of the codons do not code for amino acids at all, but are used to terminate the message and stop synthesis (the process in which amino acids are put together in polypeptide chains) of the polypeptide chain. As a result of using this three-letter code, the DNA molecule has to have three times as many building blocks as the protein for which it codes. The codons for all the different amino acids are shown in Figure 4-1.

The genetic code is almost completely universal; that is, the same three mRNA letters code for the same amino acids in bacteria and in man, and in all other species. Indeed, the whole process of **translation**, in which the DNA code is made into protein, is almost the same in all forms of life. In the translation process, the message contained in the mRNA (which was transcribed from the DNA) is translated into a precise sequence of amino acids, which in turn form the protein.

The second letter of the codons

	U		C		A		G	
U	UUU	Phe	UCU	Ser	UAU	Tyr	UGU	Cys
	UUC	Phe	UCC	Ser	UAC	Tyr	UGC	Cys
	UUA	Leu	UCA	Ser	UAA	End	UGA	End
	UUG	Leu	UGG	Ser	UAG	End	UGG	Trp
C	CUU	Leu	CCU	Pro	CAU	His	CGU	Arg
	CUC	Leu	CCC	Pro	CAC	His	CGC	Arg
	CUA	Leu	CCA	Pro	CAA	Gln	CGA	Arg
	CUG	Leu	CCG	Pro	CAG	Gln	CGG	Arg
A	AUU	Ile	ACU	Thr	AAU	Asn	AGU	Ser
	AUC	Ile	ACC	Thr	AAC	Asn	AGC	Ser
	AUA	Ile	ACA	Thr	AAA	Lys	AGA	Arg
	AUG	Met	ACG	Thr	AAG	Lys	AGG	Arg
G	GUU	Val	GCU	Ala	GAU	Asp	GGU	Gly
	GUC	Val	GCC	Ala	GAC	Asp	GGC	Gly
	GUA	Val	GCA	Ala	GAA	Glu	GGA	Gly
	GUG	Val	GCG	Ala	GAG	Glu	GGG	Gly

The first letter of the codons (5' end)

Figure 4-1 Codons for all amino acids. The codons are groups of three nucleotides found in the mRNA that specify a single amino acid. (Abbreviations for the amino acids are spelled out in the Appendix.)

The Process of Translation

Attachment of mRNA to the Ribosome

Let's find out how translation happens. First, the mRNA is made from the template (–) strand of the DNA in the region that codes for a protein. This region is called the **gene**. In this transcription process, the eukaryotic cells (more complex, higher-order cells) often produce mRNA that has extra portions of RNA (introns). These introns have to be removed before the mRNA can actually be attached to the ribosome (the large RNA protein complex where protein is made). After this is done, the mRNA is ready to be translated (Fig. 4-2).

All mRNA molecules start with an AUG **start codon** near the 5' end of the chain (Fig. 4-3). The message continues in three-letter codons until it reaches one of the **stop codons** (UAA, UAG, or UGA), which *terminates* the message.

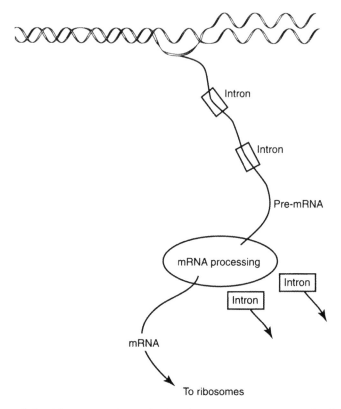

Figure 4-2 Messenger RNA (mRNA) processing. Transcribed pre-mRNA has excess material (introns) that are removed to allow the message to be correctly translated into proteins.

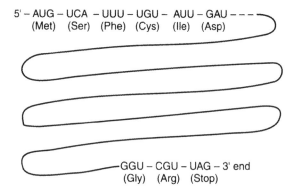

Figure 4-3 A portion of messenger RNA (mRNA) showing the start (AUG) and the stop (UAG) codon. (Abbreviations for the amino acids are spelled out in the Appendix.)

Initially, the mRNA is bound to the smaller of the two sub-units that make up the ribosome. Then, a tRNA molecule is attached, bearing its precious load of a single amino acid. As noted in Chapter 3, the transfer RNA (tRNA) has earlier been joined to this amino acid, not just to any of the 20 amino acids. This joining of an amino acid to the tRNA is a very specific reaction so that the right amino acid is on the correct tRNA). The process is accomplished by a highly selective enzyme that makes sure that both parties are right for each other.

Attachment of tRNA to the Ribosome and mRNA

As mentioned in Chapter 3, the tRNA also must be able to identify the code on the mRNA that is specific for the amino acid it carries. This takes place at the end of the tRNA most distant from the amino acid, the **anticodon** region. The anticodon region contains three nucleotides that are complementary (ie, they all make base pairs) with the three nucleotides in the codon. Thus, if the codon sequence were 5'-CAG-3', the complementary anticodon on the tRNA would be 3'-GUC-5'.

The first tRNA contains a three-letter sequence (CAU—the **anticodon**), which is complementary to the AUG start codon on the mRNA (Fig. 4-4). Methionine is the only amino acid that has "that" anticodon, so it is bound to that tRNA. This tRNA, with its specific amino acid is then hydrogen-bonded (see Chapter 1) through its anticodon to the mRNA at the AUG site. After the tRNA is in place, the other (large) ribosomal subunit is brought into place with the small subunit containing the mRNA–tRNA complex. It is the complementary **base pairing** (hydrogen bonding) between the mRNA codon and the tRNA anticodon that specifies exactly which amino acid is to be placed in each position as the polypeptide chain is elongated.

After the first tRNA finds its complementary codon on the mRNA and after both parts of the ribosome are in place, another tRNA molecule, carrying its specific amino acid, is then bound to the next codon on the mRNA. This places that tRNA at a site next to the first tRNA already on the ribosome. This tRNA must have an anticodon complementary to the next three bases on the mRNA and will be carrying the specific amino acid belonging to that anticodon. In this way, the mRNA specifies the sequence of tRNA molecules and the amino acids they carry.

By specifying the order of tRNA molecules to be placed in the proper position on the ribosome, the mRNA precisely specifies the order of amino acids in the growing chain of amino acids.

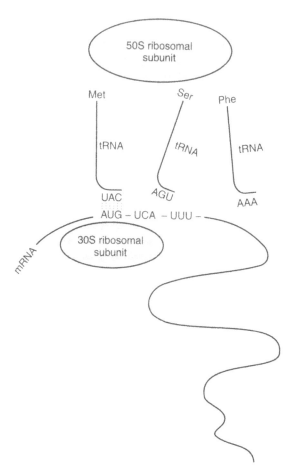

Figure 4-4 Initiation of protein biosynthesis. The transfer RNA (tRNA) carrying the first amino acid (Met) contains an anticodon (CAU), which is hydrogen-bonded to the AUG codon at the start of the mRNA on the 30S ribosomal subunit as shown. (Met = methionine; Ser = serine; Phe = phenylalanine.)

Making the Polypeptide Chain

Once the two correct amino acids are next to each other, a bond (peptide bond) is formed between the NH_3^+ (amino group) of the second amino acid and the $COOH^-$ (carboxyl) group of the first amino acid (see Fig. 2-2). The peptide bond is an especially strong bond (covalent) (Fig. 4-5).

After the peptide bond is formed, the first tRNA, which has now lost its amino acid, is displaced out of its ribosome nest and

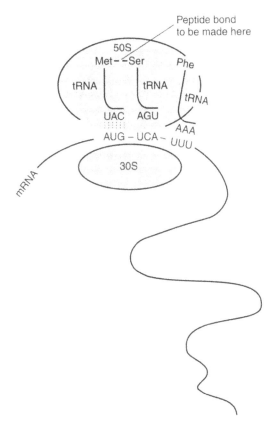

Figure 4-5 Addition of the next amino transfer RNA (tRNA)-amino acid complex and formation of the first peptide bond. After the first amino tRNA-amino acid complex is in place, the 50S ribosomal subunit is added, and an additional tRNA molecule with its attached amino acid is added in the order prescribed by the mRNA. A peptide bond is formed between the two amino acids. (Met = methionine; Ser = serine; Phe = phenylalanine.)

the tRNA carrying two joined amino acids (dipeptide) is moved over to the site where the first tRNA had been (Fig. 4-6). Then a new tRNA, which has an anticodon complementary to the next three-letter codon on the mRNA, is bound to the ribosome. Its associated amino acid is placed near the dipeptide. Another peptide bond is formed between amino acid 2 and amino acid 3, making a tripeptide, following which the vacant second tRNA molecule is displaced.

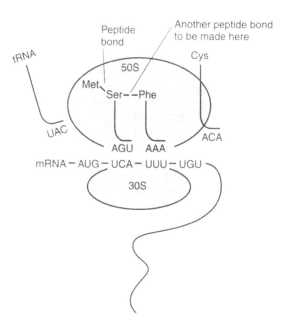

Figure 4-6 Elongation of the polypeptide chain. After formation of the first peptide bond, the initial transfer RNA (tRNA) falls off, and the next tRNA molecule moves over and an additional tRNA-amino acid is added to the vacant site. The next peptide bond is formed, and the process begins again. (Met = methionine; Ser = serine; Phe = phenylalanine; cys = cysteine)

This process is continued at a rate of up to 900 amino acids per minute until the amino acids are all attached in the order specified by the mRNA. Because the bond between the amino acids is a peptide bond, the long chain is called a **polypeptide**, as noted in Chapter 2. When folded into their final structure, polypeptide chains are known as proteins.

The End of Translation

At the end of each message, a stop codon appears and tells the ribosome to quit translating (without coding for a tRNA). Synthesis stops, and the newly formed polypeptide chain is released from the ribosome. The two ribosomal subunits then separate and prepare to start translating another mRNA.

During and shortly after the translation process (Fig. 4-7), the protein is folded into its proper secondary and tertiary structure. This folding is sometimes helped by other proteins, but mainly

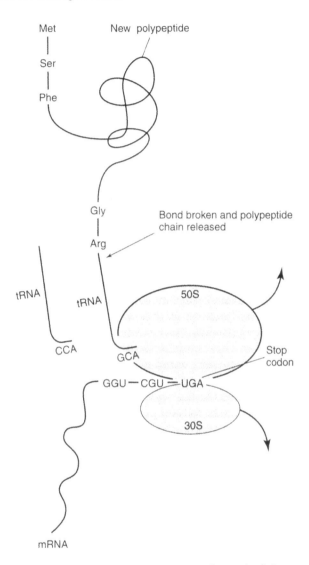

Figure 4-7   Termination of translation. At the end of the messenger RNA (mRNA), one of three codons appears (UAA, UAG, or UGA), which signals the end of the message. No transfer RNA (tRNA) contains anti-codons to these codons. They provide a signal to the ribosome to release the final tRNA and to break the bond between the tRNA and the growing polypeptide chain, as illustrated here. (Abbreviations for the amino acids are spelled out in the Appendix.)

comes about because of the amino acids that have been placed in the polypeptide chain. In the next chapter, we will see that the small changes in the order of the amino acids in the polypeptide chain can make enormous differences in protein structure and function.

REGULATION

One of the great classical works of music is Dukas' "The Sorcerer's Apprentice." As the story goes, the apprentice to the sorcerer learns how to make the brooms carry water for him, but he doesn't learn how to make them stop. A flood results and is abated only when the sorcerer returns and brings the brooms into control.

So it is with the manufacture of proteins. Cells need to **regulate** or control the number and type of proteins made. This is not only because making unnecessary proteins is energetically taxing for the cell, but also because it diverts the cell's functions from making *necessary* proteins. How is this regulation accomplished?

First, note that only a small portion of the DNA in a cell is actually used as a genetic message for mRNA. Some of the rest of the DNA is used for regulatory purposes. Generally, there is a regulatory region next to the region of the DNA used as the gene (the portion that codes for the protein). In bacteria, this entire region—the message region plus the regulatory region—is called the **operon** (Fig. 4-8).

The regulation (or control) region acts to control the amount of mRNA that is made. As previously noted, mRNA is often short-lived. So more mRNA of a given type produces more of

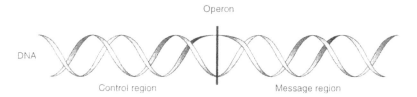

Figure 4-8   The operon is a unit of DNA that contains both a control (regulatory) and a message region. The control region is always on the 5' side of the message region. The control region regulates the amount of message (mRNA) that is transcribed from the message region of the DNA.

those proteins. By regulating the amount of mRNA produced, the cell can control the number and type of proteins needed within the cell. How does this regulation work? Several different mechanisms are at work to accomplish this. Two examples follow.

Regulatory Mechanisms

Negative Regulation

The enzyme that makes RNA from DNA, **RNA polymerase**, does not bind to the DNA at random, but carefully selects its starting point. These starting points are scattered throughout the DNA and are found in the specific regions of DNA mentioned above–**control regions**. The control region contains a site where the RNA polymerase actually binds to the DNA next to the message portion of the gene. This works like a "bookmark" to tell the RNA polymerase exactly where to begin transcribing the message. But suppose something else were already bound to the region where the RNA polymerase was supposed to bind. Then RNA polymerase would not bind and mRNA transcription would not take place (Fig. 4-9).

The protein that binds to the region where the RNA polymerase should bind is called a **repressor**, and it totally prevents transcription of the adjacent message region. If the organism needs more of the protein that is encoded by the message region, the repressor protein is pulled off that position by a molecule called an **inducer**. Once the repressor protein is removed, transcription begins, mRNA is produced, and the necessary protein is made. When enough enzyme has been made, the inducer is

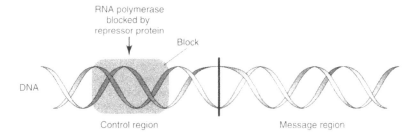

Figure 4-9 Regulation in the control region. When a repressor protein blocks the RNA polymerase binding site in the control region, RNA polymerase is unable to attach to the DNA and cannot make mRNA in the message region. No transcription occurs.

removed and the repressor returns to its position on the DNA, stopping further manufacture of the protein.

Figure 4-10 portrays the classic example of negative control mechanism. This enzyme is the **lac operon**, which controls the

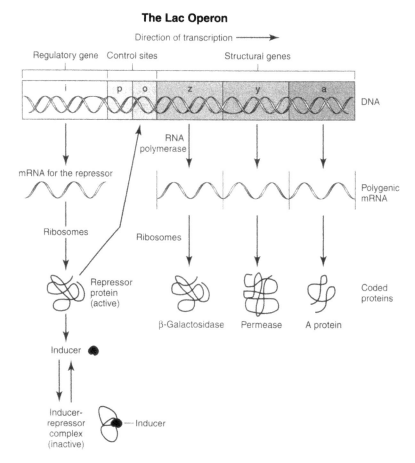

Figure 4-10 The lac operon. This is a more detailed picture of the control used in a specific case—that needed to digest milk sugar (lactose). The repressor protein binds the control region, preventing binding of RNA polymerase. No transcription occurs. When lactose is present, it acts as an inducer by binding to the repressor. With lactose bound, the repressor cannot bind the control region of the DNA, so RNA polymerase binds there and initiates transcription of the message (structural) genes. One of the enzymes made, β-galactosidase, breaks down lactose, which then cannot bind the repressor. So when lactose is present, enzymes to digest it are made. When lactose is not present, the enzymes to break it down are not made. This is an example of negative regulation.

digestion of lactose, a milk sugar. When lactose (a two-ring sugar) is present, it binds to the repressor, which causes the repressor to fall off the DNA. RNA polymerase can then bind to the site and make the mRNA. The mRNA is translated into proteins, one of which is an enzyme, β-galactosidase, which attacks lactose. Lactose is broken down into two single-ring sugars, glucose and galactose, both of which can be used by the body for energy. When lactose is broken down, it can no longer act as an inducer, so the repressor binds the DNA again. This prevents the transcription of the mRNA that produces the enzyme that breaks down lactose. This switching process is repeated as often as needed and occurs generally when lactose is present.

Some people who are lactose-intolerant have faulty regulatory mechanisms. This results in too much lactose left in the body, which can cause pain and nausea.

Although this regulatory mechanism is not used in all cases in which transcription is regulated, it illustrates the kind of **negative control** that the cell uses to make sure that only the proper amounts of protein are made (see Fig. 4-10). This level of control is called **transcriptional control**, because it occurs at the level of transcription.

Positive Control

There is another kind of regulation that is really a **positive control** mechanism. In this case, one of the proteins made when the message is transcribed makes a product that binds to an inactive repressor molecule and activates it. The activated repressor molecule then binds to the control region. This prevents the RNA polymerase from binding and thus turns off production of the product. So, in this case, the end product of the transcription regulates the transcription process.

A classic example of positive control is the trp (tryptophan) operon (Fig. 4-11). The message region contains the message for five different proteins. One of the five proteins is an enzyme needed to make tryptophan, one of the 20 amino acids. When other sources of tryptophan dry up, the tryptophan falls off the repressor and causes it to be unable to bind to the control region. The result is that RNA polymerase binds and transcribes the needed mRNA, from which the enzymes needed for making tryptophan are translated. These enzymes help make tryptophan until there is enough and a little more. Then, the excess tryptophan binds to the repressor and allows the repressor to bind to the control region of the operon once again. This in turn pre-

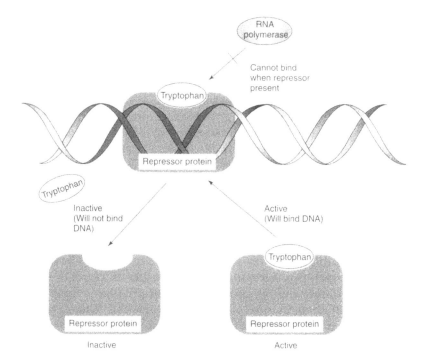

Figure 4-11   The trp operon. Excess tryptophan (trp; one of the 20 amino acids used in proteins) binds a repressor, giving it the ability to bind the control region of the operon. RNA polymerase cannot bind, and no further tryptophan is made. When there is a lack of tryptophan, the repressor does not have tryptophan around, and remains inactive as a repressor. RNA polymerase then makes the messenger (mRNA) necessary to make the enzymes that make tryptophan. This is an example of positive control.

vents RNA polymerase from binding and shuts down the manufacture of tryptophan until it is needed again.

Other Kinds of Regulation

There are many other kinds of regulation of protein synthesis, most of which fall outside the scope of our discussion. However, a very important kind of regulation is the regulation of the translation of the mRNA itself. For instance, the number or kinds of tRNA molecules can be strictly controlled, resulting in regulation of translation. The amount of ribosome production also greatly affects the amount of translation. For instance, in fast-growing cells, ribosomes are produced very rapidly and in abundant numbers. In addition, the kinds and structures of mRNA molecules also regulate protein synthesis, since some structures of mRNA are more easily "read" on the ribosome.

Regulation of the amounts and kinds of proteins in cells is very important. Much energy and effort are put into regulation mechanisms so that all cellular functions take place in the correct order. Sometimes, cells make mistakes, which are very costly to the cell and to the body. For instance, many kinds of cancer are really manifestations of the cell's inability to regulate its own affairs. Things get turned on and just don't seem to stop. The result is overproduction of cells that are often not well formed and do not function properly.

MUTATIONS

What would happen if there were a change in one or more nucleotides in the regulatory region or in the message region of the operon? A **mutation** occurs. From time to time, we see on TV grotesque humanoids derived from Hollywood's imagination. These beings, which bear little resemblance to anything living or dead, are called **mutants**. We are never told exactly what this term means. But we know that something in their genetic makeup has gone awry.

A mutation occurs when there is a change in one or more nucleotides in a gene or in the area of DNA that regulates transcription of the gene. In spite of the cells' best proofreading and repair efforts, mutations occasionally do occur. A number of factors—many yet unknown—can cause changes in sequence in the DNA. These mutations in turn may cause the mRNA to be faulty, which in turn causes the wrong amino acid(s) to be placed in a protein. Depending on which amino acid is affected, the change in the protein may not be noticed, may be damaging to the protein, or may even be lethal.

Point Mutations

A **point mutation** is a change or substitution in a single base, where, for instance, an A might be substituted for a G. Such a substitution may not be in the entire base; perhaps an atom or two are merely removed from the base. For instance, if we look at the structure of cytosine and uracil (see Fig. 4-12), we can see that by merely removing the amino group (NH_2) of cytosine and

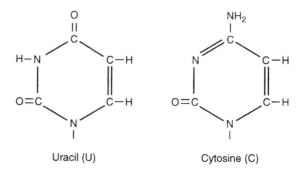

Figure 4-12   Mutation between cytosine and uracil. If some event (such as ultraviolet [UV] light) causes the amino group (NH_2) to be removed from the cytosine and it is replaced with an oxygen, uracil is formed. This often happens in nature, but uracil does not belong in DNA and must be removed by special enzymes. If uracil is left in the DNA, transcription gives codons with an A where there should have been a G and results in a mutant protein.

attaching an oxygen instead, a different base, uracil, is formed. This happens frequently all by itself. As a result, DNA has a special enzyme to check for this change (remember that uracil does not belong in DNA).

Damage to DNA is often caused by radiation or chemicals. **X-rays, gamma rays, ultraviolet light**, and various chemicals can cause mutations. No matter where we turn, some food additive or chemical that we have relied on is shown to cause mutations in the DNA. Even the way we cook our foods, such as frying or broiling at high temperatures, can make substances that cause mutations. These causative agents are called **mutagens**. The results of mutations can be mild disorders, mild chronic diseases, very severe diseases, aggravated disability, or death.

There are three different categories of point mutations: base substitution (most common), base deletion, and base insertion.

Base Substitution

Only when a mutation occurs in a **germ cell** (an egg or sperm cell) can it be transmitted from parents to children. Possibly the most common mutation is **base substitution**, in which one nucleotide is substituted for another. This produces a **missense mutation** or in some cases a **nonsense mutation** (Table 4-1).

Table 4.1 Various Kinds of Mutations

Residue number	1	2	3	4	5	6	7	8	9	10

Normal β gene A T G G T G C A C C T G A C T C C T G A G G A G A A G T C T G C C
　　　　　　　　　　Val　　His　　Leu　　Thr　　Pro　　Glu　　Glu　　Lys　　Ser　　Ala

(a) Missense mutation G T G C A C C T G A C T C C T G T G G A G A A G T C T G C C
　　　　　　　　　　　　Val　　His　　Leu　　Thr　　Pro　　Val　　Glu　　Lys　　Ser　　Ala

(b) Nonsense mutation G T G C A C C T G A C T C C T G A G G A G T A G T C T G C C
　　　　　　　　　　　　Val　　His　　Leu　　Thr　　Pro　　Glu　　Glu　　Stop

(c) Frameshift mutation G T G C A C C T G A C☐C C T G A G G A G A A G T C T G C C
by deletion　　　　　　Val　　His　　Leu　　Thr↑　　Leu　　Arg　　Arg　　Ser　　Leu
　　　　　　　　　　　　　　　　　　　　　　deletion

(d) Reversion G T G C A C C T G A C☐C C T G A G G C A G A A G T C T G C C
　　　　　　　　Val　　His　　Leu　　Thr↑　　Leu　　Arg　↑　　Lys　　Ser　　Ala
　　　　　　　　　　　　　　　　　　deletion　　　　insertion

Abbreviations for amino acids are spelled out in the Appendix.

Sickle Cell Anemia—A Genetic Disease

One of the most widely studied of the genetic diseases caused by a base substitution is **sickle cell anemia**. It affects about 1 in every 250 African Americans and normally causes death before age 30. Sickle cell anemia is a disease that makes the red blood cell look like a crescent moon (sickle cell), rather than having the plump and round shape of a normal cell (Fig. 4-13).

Within sickle cells, the protein (**hemoglobin**) that carries the oxygen from the lungs to the peripheral parts of the body has been damaged. Hemoglobin contains four separate polypeptide chains of amino acids. Two chains are identical, containing 141 amino acids each. The other two are identical and contain 146 amino acids.

In sickle cell hemoglobin, the longer chains have a different amino acid at position 6 compared with normal hemoglobin:

Normal hemoglobin

Val–His–Leu–Thr–Pro–**Glu**–Glu–Lys–

Sickle hemoglobin

Val–His–Leu–Thr–Pro–**Val**–Glu–Lys–

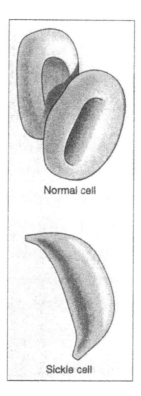

Normal cell

Sickle cell

Figure 4-13   Normal erythrocyte (red blood cell) and sickled erythrocyte. The shape change is caused by the change in a single amino acid
in the hemoglobin molecules in the sickled cells

where Val = valine, His = histidine, Leu = leucine, Thr = threonine, Pro = proline, Glu = glutamic acid, Lys = lysine (6 of the 20
common amino acids).

Thus, sickle cell anemia, which is fatal to thousands of Africans
and African Americans each year, is the result of the substitution
of a *single* amino acid in two of the four long polypeptide chains
of hemoglobin. This single change is caused by a mutation in the
gene that codes for the larger chain in hemoglobin.

From the information we have just discussed about protein
synthesis and looking at the codons shown in Figure 4-1, we can
see that in the gene, a single codon must have been changed
from either a GAA or a GAG, which codes for the amino acid
Glu, to a GUA or a GUG, which codes for the amino acid Val.
This substitution of a single nucleotide (A → U) out of the 438
used by the gene to code this polypeptide chain causes a mutation that is ultimately lethal to the individual.

This single amino acid change can alter the entire structure of the cells, as shown in Figure 4-13. How does this come about? Glutamic acid (Glu) is a negatively charged amino acid and can hydrogen-bond with water. It is replaced by an amino acid that has no charge (Val) and is very hydrophobic (water-hating). Then at position 6, instead of a charged, water-loving side chain, there is a water-hating side chain. This region of the polypeptide chain is exposed to the water in the cell and, because of the change, now tries to hide from water. It does this by binding with a neighboring hemoglobin molecule that has a similar hydrophobic patch. The result is that the hemoglobin molecules gather together in large bunches and can't carry oxygen well at all. So the person who has this disease literally suffocates, because he or she cannot get oxygen out to the cells in the body.

Other Genetic Disorders

Sickle cell anemia is but one of many genetic disorders. In some of these disorders, the causes are known, such as in sickle cell anemia. Some have been mapped to specific parts of the human **genome** (human genetic material found in the chromosomes in all human cells). The genes for other diseases have now been identified as well. For instance, cystic fibrosis, a devastating disease that afflicts 30,000 children and young adults in the United States alone, is caused by a genetic disorder. The gene responsible for this disorder has now been found on chromosome 7 in the human genome. Genes responsible for many other genetic disorders have also been identified (Table 4-2).

With almost any protein in any cell, the same scenario can be repeated. By changing a single amino acid, the characteristics of the active sites of enzymes may be altered. In other cases, the structure of the entire protein might be changed. The result is that the mutation in the DNA may cause nonfunctional or only partially functional proteins to be made.

In addition to mutations in the gene, mutations may occur in the region of the DNA that regulates the transcription of the gene. In these cases, the primary sequence of the protein itself may remain unchanged, but either too little or perhaps too much protein is made. This causes another set of problems.

Base Insertion and Base Deletion

Another kind of mutation is that in which a nucleotide is deleted from the gene, or a new nucleotide is inserted. It is possible that a mistake is made while the DNA itself is initially replicated and

TABLE 4.2 Common Genetic Diseases*

Inborn Errors of Metabolism	Approximate Incidence Among Live Births
Cystic fibrosis (mutated gene unknown)	1/1600 whites
Duchenne muscular dystrophy (mutated gene unknown)	1/3000 boys (X-linked)
Gaucher's disease (defective glucocerebrosidase)	1/2500 Ashkenzi Jews, 1/75,000 others
Tay-Sachs disease (defective hexosaminidase A)	1/3500 Ashkenazi Jews,1/35,000 otners
Essential pentosuria (benign condition)	1/2000 Ashkenazi Jews, 1/50,000 others
Classic hemophilia (defective clotting factor VIII)	1/10,000 boys (X-linked)
Phenylketonuria (defective phenylalanine hydroxylase	1/5000 among Celtic Irish, 1/15,000 others
Cystinuria (mutated gene unknown)	1/15,000
Metachromatic leukodystrophy (defective arylsulfatase A)	1/40,000
Galactosemia (defective galactose-1-phosphate uridyl transferase)	1/40,000
Hemoglobinopathies	**Approximate Incidence Among Live Births**
Sickle cell anemia (defective β-globin chain)	1/400 US blacks; in some West African populations the incidence of heterozygotes is 2/5
β-thalassemia (defective β-globin chain)	1/400 among some Mediterranean populations

*Although most of the over 500 recognized recessive genetic diseases are extremely rare, in combination they represent an enormous burden of human suffering. as is consistent with mendelian mutations, the incidence of some of these diseases is much higher in certain racial groups in others.

a nucleotide is left out or an extra one inserted. Such mistakes do occur, but only rarely (about 1 in 10 billion). However, even such a low rate of error has a measurable effect on the mutations that occur.

Insertion or deletion mutants are almost always lethal, because they cause what is known as a **frame-shift mutation**. Recall how the genetic code is set up. There are codons that each contain three letters, all next to one another with no spacer between them. So a deletion or addition of a single nucleotide will change every codon behind it. As an example, look at this sequence in mRNA:

-AAA AGC ACU CCG CGA UUC-

giving the amino acid sequence:

-Lys-Ser-Thr-Pro-Arg-Phe-

where Lys = lysine, Ser = serine, Thr = threonine, Pro = proline, Arg = arginine, Phe = phenylalanine.

If we delete the fourth A, we have

-AAA GCA CUC CGC GAU UC-

which gives us:

-Lys-Ala-Leu-Arg-Asp-Ser-

where Lys = lysine, Ala = alanine, Leu = leucine, Arg = arginine, Asp = aspartic acid, Ser = serine. This creates an entirely different sequence.

Not only is the second amino acid altered, but all of those that follow are altered as well. Almost certainly such a mutation would lead to a protein that will be enormously different from the native structure and this often causes death.

However, there can also be a **reversion**, in which a base substitution returns to the original base, or an insertion is compensated for by a base deletion nearby, or vice versa. In the latter cases, the frame shift is reversed, and the rest of the protein is like the native protein. This often allows the organisms to continue with an almost normal life cycle.

Neutral Mutations

Some mutations change only a single nucleotide into another, but the amino acid for which it will code remains the same. For instance, CTA might be changed to CTG, but both of these code for Leu. These changes are then *silent*. If a mutation occurs in which one amino acid is substituted for another but does not affect protein activity, it is called a **neutral substitution** or **neutral mutation**.

From the examples in Table 4-1, we can see that some mutations may be very harmful. However, a beneficial side to mutations should be mentioned. In one-celled organisms and higher forms of life, the environment around them changes and sometimes becomes so harsh or changes so much that the normal bacteria or organisms cannot live. Often mutants have been formed that seem able to resist the negative effects of the new environment and can then live in the new surroundings. Sometimes the numbers and kinds of mutations increase as a result of the environmental pressures. For these reasons, it is good to have some mutations. Of course, one of the problems stemming from mutations is that some organisms are becoming resistant to antibiotics and other drugs.

SUMMARY

Genetic information that is stored in the chromosome is used to make proteins of any type or variety that the cell needs. Once the need is apparent, the cell calls on the DNA to produce a piece of mRNA containing the necessary information. This mRNA is then transported to the ribosome, where the tRNA molecules bearing correct amino acids bind to it in accordance with the genetic message contained in the mRNA. Peptide bonds are formed between adjacent amino acids, producing a long chain of amino acids, the polypeptide chain. This chain then folds in a specific manner, and a protein is born!

The manner in which the amount of protein is regulated is very complex. There can be either a positive or a negative control system, or perhaps other types. It is important to realize that the translating system is not always making all the proteins in a cell, but is limited to make those that the cell needs at a particular time.

There are two fundamental ways in which the manufacture of proteins in cells and organisms might be damaged. First, mutations may be present which cause the proteins that are made to be unsuitable for their purpose. In these cases, severe disease or death may result. A mutation may occur in the control region, or there may be a mutation in the repressor molecule. Under these circumstances, the control of transcription is damaged, deregulating the cell. Such can result in diseases like cancer or perhaps diseases in which certain proteins are lacking altogether. In either the case of mutation of the proteins, or damage of the control mechanisms, severe problems can result.

5

ALTERING GENETIC MATERIAL IN BACTERIA

WHAT YOU WILL LEARN IN THIS CHAPTER

- The composition of bacteria
- How bacteria exchange plasmids and chromosomal material naturally
- How plasmids can be extracted from bacteria and isolated
- How new pieces of DNA can be inserted into old plasmids by "cutting" and "pasting"
- How plasmids containing new DNA can be put back into bacteria

---

Up to this point we have learned about the way in which nucleic acids and proteins are made and how cells regulate these functions. Our real purpose in this book is to discuss how to change the genetic information that will result in changes in the proteins that are made or in the way a cell operates. Most often, this effort is directed toward changing a certain protein to make it functional.

It is important that we start by making genetic changes in bacteria, because they are fairly simple living organisms. Bacteria are part of a large group of organisms called **prokaryotes** (single-celled organisms that don't have nuclei). We discuss bacteria and higher-order cells here because their functions are relevant in genetic engineering. (However, a complete discussion of all their functions would push us well outside the scope of this book.) In later chapters, we will discuss more complex cells, such as those from plants and animals and those from humans. These cells are called **eukaryotes**.

WHAT ARE BACTERIA?

Bacteria are tiny living organisms of different sizes and shapes. They are found everywhere. Television ads often call them germs, although this term is also applied to viruses. Bacteria are single-celled structures, which are entirely capable of maintaining and reproducing themselves. They can be found living under the most difficult circumstances, such as in the hot ponds in Yellowstone Park and in thermal chimneys deep in the ocean, as well as in the cold glaciers of the mountains. Bacteria live on almost any kind of food and adapt well to diverse environments, including oil spills and mine tailings.

Although these ubiquitous little creatures cause illness from time to time, bacteria are extremely useful to mankind and essential for our well-being. For instance, bacteria help us extract energy make building blocks from the food we eat. Without those bacteria, we would not be able to use a lot of the food we eat. We need to look at them more closely, not only because they are important for life, but because they are especially important for genetic engineering purposes.

Bacteria contain all the components and machinery necessary for sustaining life and for reproducing themselves. They function in a much less complex manner than we do, but the method is there. So, bacteria are often used as the windows through which we observe the patterns of living things. Moreover, they grow fast (with progeny born every 20 minutes under good conditions) and economically, and, as yet, no one has loudly complained about opening up and studying these creatures and using them to find out what life is all about.

A typical bacterium is shown in Figure 1-4. Certain portions are labeled to show some of the organelles (specialized cells) necessary for the sustaining and continuation of life. In bacteria, the blueprint of life—DNA—is found in the **chromosome**. As previously discussed, this genetic information is transferred through a message to the ribosome machinery in which proteins are manufactured according to the exacting specifications contained in the DNA blueprints.

Chromosomes and Plasmids

Let's talk a bit more about the chromosomal material of bacteria. Not much protein is associated with this DNA, certainly not as much as in eukaryotic cells. Most of the time in bacteria, the DNA is found in a large circular, double-stranded form. The DNA in the chromosome contains a number of segments called **genes**, each of which contain the code for a particular protein. For this reason, the chromosomal material in a cell is often called the **genome** (gene + chromosome). These genes within the genome were referred to in the last chapter as "message regions" of the DNA.

In addition to the main chromosome, another type of DNA is often found in bacteria. This DNA is extremely useful for the well-being of the cells. It comes in small, circular structures called **plasmids** (sometimes dubbed **minichromosomes**). Plasmids contain certain highly specialized genes. Specifically, the sex factor of bacteria is found in plasmids and also genes that code for antibiotic resistance. Plasmids are of great interest to us, because they are fairly simple in structure and easy to extract from bacteria. Probably more important is the fact that when plasmids are isolated, we can alter them as needed (as we will show) and then reinsert them into the bacteria. The bacteria then treat these reinserted plasmids as part of themselves, and the plasmids are duplicated as the cells divide.

As already noted, within some plasmids are regions of genes that give bacteria resistance to certain antibiotics. So if we were to look closely at a plasmid, we might find the organization shown in Figure 5-1, with areas that provide resistance to ampicillin (Ap^r), and tetracycline (Tc^r).

The plasmid is an ideal structure for genetic engineers for two reasons: (1) it contains genetic information used by the bacteria, and (2) the plasmid itself is not essential to bacterial functions. So, it is possible to manipulate this DNA without upsetting the bacteria.

BRINGING NEW DNA INTO OLD BACTERIA

How is manipulation of genetic material in a bacterium to be done? Sometimes it happens naturally. For instance, from time to time bacterial cells go through a process called **conjugation**, in which they share genetic material (Fig. 5-2). They become attached through a **pilus** (a hollow tube that connects the cyto-

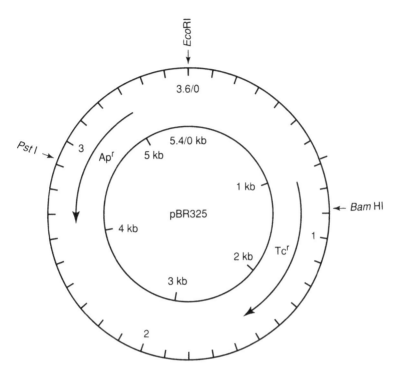

Figure 5-1 The genetic map of a plasmid showing sites where some of the restriction enzymes (e.g., *Eco*RI, *Pst*I and *Bam* HI, discussed later in this chapter) cut the plasmid. The arrows with labels Tc^r and Ap^r indicate the regions that provide the bacteria with resistance to tetracycline and ampicillin. This plasmid has 5400 bases (kb = kilobases, thousands of bases).

plasm of the two bacteria), and DNA is transferred from one bacterium to the other. During the conjugation process, a full plasmid or merely a piece of it may be transferred. In this way, bacteria can exchange resistance to antibiotics and other traits that will help them withstand the rigors of their environment. At times, even portions of the bacterial chromosomes are transferred as well.

In addition to conjugation, plasmids can sometimes be inserted into bacteria just by being present in the culture medium. This is called **plasmid transfer**. These plasmids move into the bacteria through pores in the membranes. Some bacteria are able to do this more readily than others. Once inside, the plasmids sometimes recombine with the bacterial chromosome. At some later time, the plasmid may be expelled from the DNA

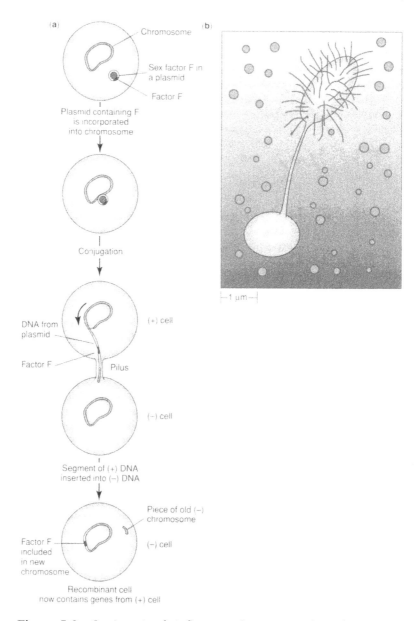

(a)

Chromosome

Sex factor F in
a plasmid

Factor F

Plasmid containing F
is incorporated
into chromosome

Conjugation

DNA from
plasmid

(+) cell

Factor F

Pilus

(−) cell

Segment of (+) DNA
inserted into (−) DNA

Piece of old (−)
chromosome

Factor F
included
in new
chromosome

(−) cell

Recombinant cell
now contains genes from (+) cell

(b)

⊢1 µm⊣

Figure 5-2   Conjugation briefly mates bacteria, and, in the process,
DNA from one bacterium is transferred to the other, as shown. Note
that a sex factor in a plasmid is transferred into a bacterium not containing that factor. In this same way, antibiotic resistance genes as well
as others can be transferred from one bacterium to another.

and form a small circular plasmid again. As it leaves, it may carry with it a portion of the cellular chromosome, or it may leave a bit of itself behind in the chromosome. In either case, the genetic information of both the chromosome and the plasmid is altered.

These kinds of genetic alterations give bacteria the ability to adapt to environmental stresses and changes, which give rise to the wide variety of bacteria found. In cases of antibiotic resistance, the alterations are acts of self-preservation for bacteria, but may present a problem for us. This is becoming increasingly apparent as more varieties of antibiotic-resistant bacterial strains are developing, which make diseases they cause harder to treat (e.g., tuberculosis)

THE TOOLS OF GENETIC ENGINEERING

Whenever free DNA is taken into bacteria from the surrounding environment, **transformation** occurs. The ability to take in DNA through the membranes varies between bacteria. Those that readily allow DNA to enter are called **competent cells**. In nature, exchange of genetic information occurs continually, through both conjugation and transformation.

Genetic engineering of a bacterium takes place when a bacterial chromosome or plasmid is changed by design. In this process, we actually decide what kinds of changes we wish to make and then go about the process of making them. If we can take some DNA from plasmids or other bacterial chromosomes and modify it in a known manner and then put that DNA inside the cell's DNA, we have then caused a change to occur in the cell's DNA. The easiest way to do this is to use one of the natural methods of transformation—plasmid transfer, as illustrated in Figure 5-3.

In order to engineer changes in DNA and insert the changed DNA into bacteria, there are several things we need to learn to do. The overall objective at this stage is to place a piece of new DNA in bacteria in order to have the bacteria make a protein that they would not usually make. So we need to do the following:

1. Extract plasmids from bacteria.
2. Cut the plasmids open in specific regions.
3. Insert a piece of DNA into the plasmid.

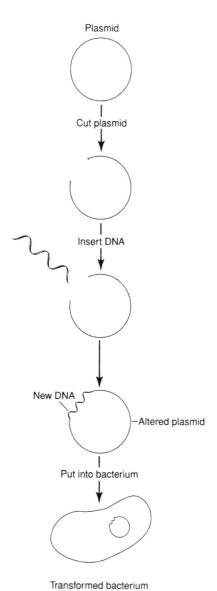

Figure 5-3 shows labels: Plasmid, Cut plasmid, Insert DNA, New DNA, Altered plasmid, Put into bacterium, Transformed bacterium

Figure 5-3 Diagrammatic overview of placing new DNA into a bacterium. First, the plasmid is opened at a specific site; new DNA is inserted into the gap and then bound to the plasmid. The altered plasmid is then placed into bacteria. The addition of the new DNA transforms the bacteria into bacteria having different characteristics.

4. Fuse (bond) the new DNA to the old DNA and close the plasmid circle.
5. Insert the plasmid into the bacteria and have the bacteria make the new protein.

It sounds complicated at first, but some neat tools have been developed to help us do this well.

Electrophoresis

If we are going to cut, insert, and glue pieces of DNA to one another, we need to be able to "see" the DNA. Although we might try to look at the DNA with an electron microscope both before and after the DNA was cut, this method is slow and technically difficult and really does not have sufficient resolution to tell us if we cut the DNA at the right place. However, a neat technique has been developed by which DNA pieces can be "seen." In this technique, we don't look at the pieces directly, but we can tell how long they are. This method is called **electrophoresis**.

Electrophoresis is one of the most useful techniques in all of genetic engineering. It uses electrical current to separate pieces of DNA having different lengths.

To use this technique, we first make a **polymerized acrylamide gel** slab by pouring a solution of acrylamide between two glass plates. The edges of the plates have been sealed so the acrylamide solution doesn't leak out before it sets, like gelatin. **Acrylamide** is a small molecule, which, when put in the right solution, attaches to others like itself, making a long chain (polymer) of acrylamide molecules. This is commonly called polyacrylamide. Before it polymerizes, acrylamide is much like any other liquid solution. After polymerization occurs, it makes a microscopic mesh of polyacrylamide fibers. Before it sets, we place a special **comb** in the top to give notches in the top of the polyacrylamide slab, which ends up looking like a castle wall. The notches are called **wells**, and this is where we put our samples (Fig. 5-4).

The glass plates with the polyacrylamide gel slab between are now placed in an apparatus that contains an upper and lower reservoir to hold salt solutions that transmit electrical current from a power supply (Fig. 5-5). Each DNA sample is placed carefully in a separate well. Remembering that DNA is negatively charged owing to all of the phosphate groups in the sugar-phosphate backbone, we put the positive electrode in the lower reservoir. When electrical voltage is applied, it causes the nega-

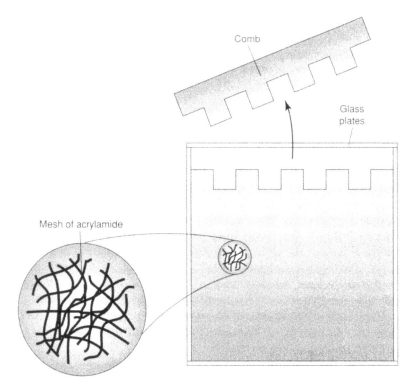

Figure 5-4 Diagram showing how polyacrylamide gels are made. Two glass plates are slightly separated, and a solution of acrylamide is poured between them. The solution polymerizes (like gelatin), forming a mesh-like structure between the plates. A comb is used to create wells into which DNA samples can be placed. These gels are used to separate DNA pieces of different length. Smaller pieces move through the gel faster because the longer pieces of DNA get tangled in the acrylamide gel strands more readily than the smaller pieces do.

tively charged DNA to be pulled down through the gel toward the positive electrode.

Because the gel is a microscopic mesh, the larger pieces of DNA move more slowly than the smaller ones because they keep bumping into the polymerized acrylamide on the way through. The result is that the DNA fragments are separated, with the smallest fragments moving the greatest distance.

Let's take a sample of whole DNA (plasmids work fine) and then the same DNA sample that has been cut with an enzyme.

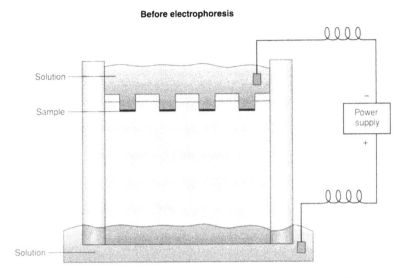

Figure 5-5 A polyacrylamide gel electrophoresis apparatus. The gel is still between two glass plates, which are then placed in a container that allows a solution to be on the top and bottom of the gel. Electrodes are placed in the solution, and the DNA sample is placed in the wells. The electrical potential between the top and the bottom pulls the negatively charged DNA through the gel to the positive electrode.

Let's place these samples in adjacent wells in an electrophoresis apparatus. Then we turn on the power supply and allow the voltage to pull the DNA through the gel. This would allow us to "see" how many fragments of DNA were made upon cutting the DNA with the enzyme.

Stain Technique

How do we "see" the DNA fragments in the gel? Actually, there are several ways to do this. The most common method is to use a stain, which colors the DNA and not the gel. This may give us a pattern that would look like that in Figure 5-6.

Radioactive Technique

Another very useful way to see DNA is to use DNA that has been made slightly radioactive. This can be done readily using a special enzyme that substitutes a radioactive phosphorus (^{32}P) for the phosphorus on the 5′ end of any DNA. Then, the radio-

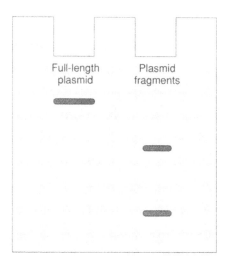

Figure 5-6   Diagram of a pattern from a plasmid that might be found after cutting the plasmid, electrophoresis, and staining. Note that the plasmid fragments are both shorter than the plasmid itself and will move faster through the gel, with the shortest fragment being closest to the bottom of the gel. After electrophoresis, the gel is removed from the glass plates, and the gel is soaked in a solution containing stain. The ink-like stain binds only to the DNA and not to the gel itself, so the DNA shows up as dark bands.

actively labeled DNA is put in the slots in the gel, and the power supply is turned on. After a period of time, during which the DNA moves through the microscopic mesh of polyacrylamide fibrils, the polyacrylamide gel slab is removed from between the glass plates and placed on a piece of x-ray film. The film with the gel next to it (generally with a piece of plastic wrap material between) is allowed to remain in the dark for several hours to expose the x-ray film. Wherever radioactivity is present, the x-ray film is exposed, leaving a black spot. So the same kind of pattern appears a; that which appears on the stained gel (Fig. 5-7).

The electrophoresis technique gives us size information on the DNA fragments produced with our scissors. These enzyme scissors are very specific for certain places on the DNA and will cut DNA nowhere else. They are called **restriction endonucleases**. The term "restriction" comes from their being restricted to

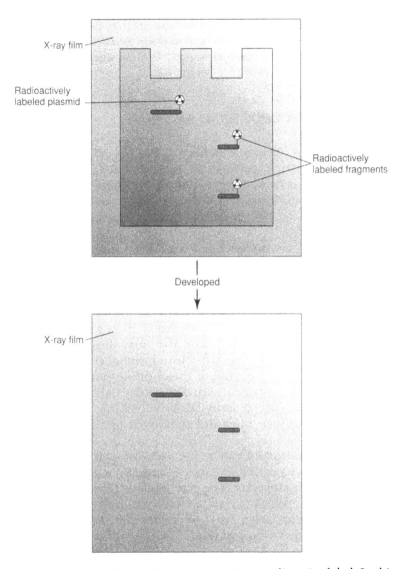

Figure 5-7   Visualizing DNA patterns using a radioactive label. In this case, the DNA is labeled with a radioactive tracer (normally ^{32}P), which travels with the plasmid or the fragments of the plasmid. After electrophoresis, the gel is removed from the glass plates, and the gel is then placed next to x-ray film in a holder and allowed to expose the film in the dark for a few hours. The film then shows the places where the DNA is in the gel. This technique is much more sensitive (that is, can "see" much smaller amounts of DNA) than staining.

specific places on the DNA. "Endo-" means they can cut the DNA anywhere between the ends (see section on Cutting Plasmids Open). Now that we have this neat tool at our disposal, let's return to the bacteria.

OBTAINING PLASMIDS FROM BACTERIA

We need to grow bacteria that have usable plasmids in them. We can obtain starting cultures of special bacteria containing specific plasmids from commercial supply houses. We can then grow these bacteria in a culture just like wild-type bacteria. They can then be harvested and broken open to release the plasmids.

Once the cells are broken open, the DNA is separated from the rest of the cell material, following which a solution of it is layered on top of a dense solution in a centrifuge tube. (A **centrifuge** is a machine that can spin a rotor, which holds centrifuge tubes, at high speeds and cause the heavy material to go to the bottom of the tube.) This tube is spun at high rates of speed in the centrifuge for several hours, during which the plasmid DNA separates from the chromosomal DNA because its density is different from that of the chromosomal DNA. These bands of DNA can be seen with ultraviolet light. We can then extract the band of plasmid DNA by poking a needle into the plastic centrifuge tube and sucking it out with a syringe (Fig. 5-8).

The plasmid obtained can be precipitated (settled) out of the dense solution by using ethanol (alcohol), which causes all the DNA to clomp together. The precipitate is then redissolved in the appropriate solution, and we have the desired plasmids in solution. There are other ways to isolate plasmids, but for our purposes this one works just fine.

Cutting Plasmids Open

Next, we need to find a pair of scissors to cut the plasmid exactly where we want. We have already mentioned these scissors before. They are restriction endonucleases. The requirements are that the scissors must be very, very small, since we need to cut the phosphodiester bond (see Chapter 3) at a specific location. These tiny scissors have to have "eyes" of their own and actually pick the sites they will cut to allow us to distinguish individual nucleotides.

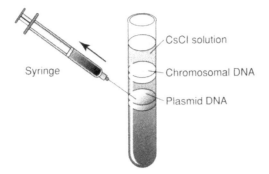

Figure 5-8   Separation of plasmid and chromosomal DNA using the centrifuge. A dense solution of cesium chloride (CsCl) is placed in a centrifuge tube, and the sample containing DNA is placed on the top, after which the tube is spun for several hours. Chromosomal DNA is less dense than plasmid DNA, so they separate into two separate bands, which can be seen in ultraviolet (UV) light. The tube is then punctured in the side with a syringe to remove the pure plasmid DNA.

For instance, suppose one small section of the DNA looked like this:

–AGGCTGGAATTCCGCTTA–
–TCCGACCTTAAGGCGAAT–

And suppose that we wanted to split it right between the two "Ts" next to each other on the upper strand and the two neighboring "As" on the lower strand.

Mother Nature has come to the rescue. It turns out that one of the defense mechanisms that bacteria have against viral infection is a special set of enzymes that look for and cut invading, nonbacterial DNA in very specific places. These enzymes, the restriction endonucleases, were discovered in the 1950s and isolated in the 1960s and later. Exactly how these enzymes find and cut these foreign DNA molelcules is something to discuss another day. However, over 700 kinds of restriction endonucleases have been discovered, and more are still being discovered. Each of these enzymes cuts DNA specifically at a unique site, which depends entirely on a particular short sequence of nucleotides in the DNA. Let's look at two examples.

EcoRI

The first restriction endonuclease that was discovered is called **Eco**RI. This enzyme was discovered in *Escherichia coli* (a bacterium); hence, its name. *Eco*RI specifically looks for regions (**recognition sites**) in DNA that have the sequence:

$$5'-G-A-A-T-T-C-3'$$
$$3'-C-T-T-A-A-G-5'$$

Once the recognition site has been found, the enzyme binds to that region and then cleaves (splits) the DNA in a very specific way. The arrows in Figure 5-9*a* show where the cuts will be made. Note that the resulting DNA fragments have ends that are **overlapping**, as shown in Figure 5-9*b*.

Often these overlapping ends are called **sticky ends**, because they tend to hydrogen-bond to their counterparts really well. Many of the restriction endonucleases make these sticky ends, but the number and sequences of nucleotides in the sticky ends are different, depending on which restriction endonuclease was used to cut the DNA.

Hpa I

The other endonuclease that we will use for an example is called **Hpa** I. It will only cleave DNA that has the following sequence:

$$5'-G-T-T-A-A-C-3'$$
$$3'-C-A-A-T-T-G-5'$$

This enzyme recognizes only the sequence shown and binds to that region of any DNA. Then cleavage occurs. Note that in this case, the product does not leave overlapping ends, but gives **blunt** or **flush ends**. See Figure 5-10 for an illustration of what a small piece of DNA cut with *Hpa*I would look like.

Many other restriction endonucleases also give blunt ends, each cleaving at a specific site having a different sequence. All the endonucleases that we will discuss will give either sticky ends or blunt ends. Since there are over 1000 different enzymes to choose from (you can buy many from a biochemical supply company), you can choose just about any particular sequence of DNA you want as a potential cleavage site and find an enzyme to cleave it there.

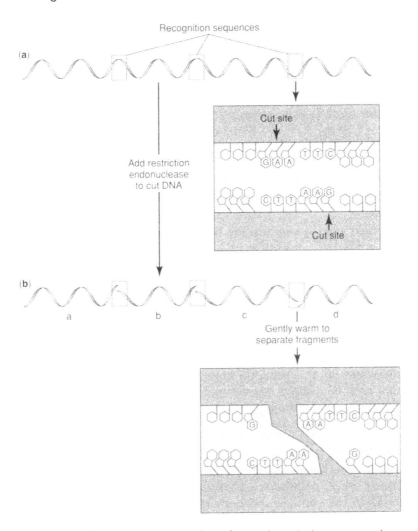

Figure 5-9 How restriction endonucleases (restriction enzymes) cut DNA. (**a**) A piece of DNA containing three recognition sites for a restriction enzyme. These are the blocks on the ribbon. A restriction enzyme binds only where the specific sequences it recognizes occur. In this case, it has to be an GAATTC, as shown. (**b**) Once the restriction enzyme binds the recognition site, it cleaves between two specific nucleotides—in this case between G-A. This fragments the DNA at this site, leaving overlapping (or sticky) ends. Each restriction enzyme recognizes and cleaves DNA in only those places that contain a specific sequence of nucleotides that the restriction enzyme recognizes.

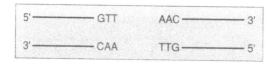

Figure 5-10   An example of a piece of DNA cleaved by a restriction enzyme (*Hpa* I) that recognizes the sequence GTTAAC and cleaves between the last T and the first A (in a 5' to 3' direction). The product is DNA that has blunt ends.

INSERTING NEW DNA INTO PLASMIDS

The next step in engineering DNA is to put a new piece of DNA into a plasmid and put it into working order again. But where do we get the new DNA? Possibly from some other organism. But there are limitations. Or, we could make the new DNA chemically. It is possible to buy chemically synthesized DNA pieces that contain up to 100 nucleotides. These pieces of DNA are made from individual nucleotides and coupled in an instrument that forms the phosphodiester bonds between the nucleotides. These nucleotides are placed in the sequence that you want. So, it is possible to go to biochemical supply companies and buy short segments of DNA having any sequence. Therefore, depending on what we wish to do, we must either obtain DNA from another source or synthesize the pieces we need.

If we are going to cut a piece of DNA from another source, we will have to know enough about that source to know the location of the appropriate restriction enzyme cleavage sites around the DNA we want. Suppose we knew that a portion of the DNA looked like that shown in Figure 5-11. By using that restriction endonuclease (*Eco*RI), we would get the piece of DNA that we want and would know its size. So by running the DNA fragments on an electrophoresis gel and using commercially available size markers, we should be able to identify the fragment we want (Fig. 5-12).

When we have identified the band (fragment) we want, the DNA can be removed from the gel by finely mincing the gel, dissolving the acrylamide, and then precipitating the DNA out of the solution using ethanol. In this way, we can purify a particular band of DNA.

The isolated DNA could be a gene for a protein, but for the moment it may be just any piece with a given sequence. If we

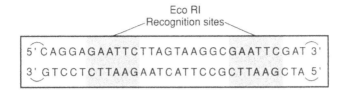

Eco RI
Recognition sites

5' C A G G A G A A T T C T T A G T A A G G C G A A T T C G A T 3'
3' G T C C T C T T A A G A A T C A T T C C G C T T A A G C T A 5'

Figure 5-11 Example using restriction enzymes to isolate a piece of DNA. Two *Eco*RI sites flank the fragment of DNA to be isolated. By cleaving the DNA with *Eco*RI, the desired fragment of DNA is removed from the longer piece of DNA (only a portion of which is diagrammed). Using gel electrophoresis, the desired fragment can then be separated from the rest of the DNA. See text for details.

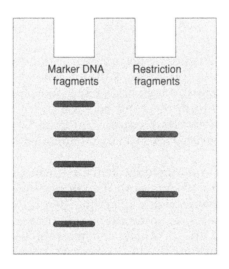

Marker DNA fragments   Restriction fragments

Figure 5-12 Electrophoresis pattern of fragments of DNA obtained by digestion of DNA using a restriction enzyme. In the left-hand lane is a set of marker DNA pieces, obtained commercially, which have known lengths. By comparing the unknown restriction fragments with the DNA markers, fragments of the size expected can be identified and then isolated by extracting them from the gel.

have isolated this piece of DNA, using *Eco*RI, we would have the following structure:

A A T T C T T A G T A A G G C C
G A A T C A T T C C G G T T A A

GLUING GENES TOGETHER

How do we attach this piece of DNA to another piece of DNA; or better yet, insert it into a plasmid? When we insert a piece of DNA into a plasmid, it is called an **insert**. One of the important considerations is that both the insert and the plasmid pieces have ends that are compatible. For instance, it would be difficult to attach an insert containing blunt ends to a plasmid containing sticky ends. The easiest way to see that the ends fit together properly is to use the same restriction endonucleases to cut out the insert and to cut the plasmid. Then we will have correct ends on both pieces.

We now place both pieces of DNA into the same vial and then incubate for a short period. The sticky ends hydrogen-bond (**anneal**) together, and the new piece of DNA is inserted into the plasmid (Fig. 5-13).

Even though the sticky ends are hydrogen-bonded, the covalent phosphodiester bonds are still not formed. So we add an enzyme, called **ligase**, which makes new phosphodiester bonds. Ligase is able to form the bond between the 3′ and 5′ ends of adjacent sugar molecules, as shown in Figure 5-13. The product is called a **chimeric DNA**, because it contains DNA from two sources.

All this works as outlined in theory, but when we actually do the experiment by placing both the cut plasmid and the DNA insert into the same vial and adding ligase, we can have a number of products. In some cases, our DNA will be inserted as we wanted it to; in others, the plasmid will reattach its own ends. Sometimes the new pieces of DNA will attach to themselves.

Initially, we won't know which of the possible end-products have occurred. It could be anything from the original plasmid reformed to a multiple insert placed in the plasmid. It will ultimately be necessary to separate these products, but it is much easier to do it later, after the plasmids have been put into bacteria. For now, we just need to realize that many kinds of products are possible.

If we use blunt-end restriction endonucleases for the experiment, the game is much the same, but we don't need to worry about the overlapping ends. However, the ligation (tying together) of the blunt ends generally takes more time and material and often is less successful.

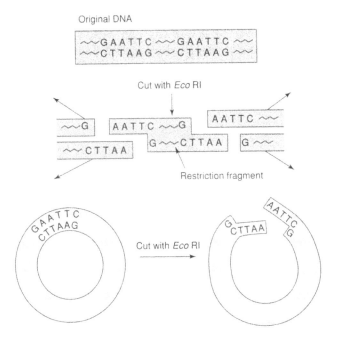

Original DNA

Cut with *Eco* RI

Restriction fragment

Cut with *Eco* RI

Put cut plasmid with restriction fragment

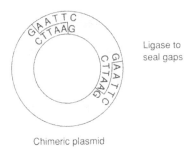

Ligase to
seal gaps

Chimeric plasmid

Figure 5-13. Diagram of cutting a plasmid and inserting a new piece of DNA. The DNA on the right could be the fragment we obtained by cleavage, as shown in Figure 5-11. By cutting the plasmid with the same restriction enzyme, similar ends are formed in the plasmid that are found in the fragment. By placing both the fragment and the cut plasmid into a vial and allowing it to mix (incubate) for a while, the plasmid ends attach to the fragment as shown. New phosphodiester bonds between the fragment and the plasmid are formed by using another enzyme, ligase. See text for details.

A final comment on the use of plasmids for use in bringing new DNA into bacteria. Plasmids can only "carry" so much DNA. If they become too large, the bacteria do not use them and the plasmids will not be reproduced properly. So there is a limit to the size of the insert that we can put into a plasmid.

SUMMARY

Bacteria are simple forms of life, grow fast, and are easy to manipulate; yet they contain the essential ingredients for genetic experiments because they have chromosomes and plasmids. They are excellent living systems to use for experimental purposes. Plasmids are tiny circles of DNA that can be transferred between bacteria either by the mating of bacteria (conjugation) or by plasmids being "picked up" from the surrounding medium by bacteria (transformation). When either of these processes takes place, the genetic information in bacteria is altered from the original state. We would like bacteria to pick up pieces for DNA that we want them to have. So, we have to be able to extract plasmids from bacteria, cut the plasmids open, insert a new piece of DNA, glue the plasmids back together, and put them back into bacteria. To "see" what we are doing, we use gel electrophoresis. Various enzymes are needed to help us. Restriction endonucleases cut DNA at specific sites, often giving sticky ends. Ligase is the enzyme that glues two pieces of DNA together. The reconstructed plasmid then can be inserted into cells giving them and their progeny new genetic information.

6

GENETICALLY ENGINEERING BACTERIA

WHAT YOU WILL LEARN IN THIS CHAPTER

- How to find the sequence of nucleotides in a piece of DNA
- How to insert a piece of new DNA at a specific location in a plasmid
- How to carry out colony screening and Southern blotting techniques
- How to identify, by screening, the genetically engineered bacteria that carry the insert of new DNA
- How to make gene libraries

With most of the tools we need in hand, we are now prepared to genetically engineer bacteria. The most direct way to do this is to isolate plasmids, insert the new piece of DNA into the plasmids, and insert the plasmids into the bacteria. We do this by combining most of the techniques we have learned thus far (see Chapter 5). In addition, we need to be able to learn the sequence of the DNA with which we are working.

FINDING THE SEQUENCE OF DNA

We're finally close to doing what we aimed to do when we started this book—**genetic engineering**. We will start with

103

bacteria in this chapter, and then move on to other living things in the chapters to come. But we still have to discuss one more tool—**sequencing DNA**. Sequencing DNA is the technique in which we identify the sequence of nucleotides in the DNA strand. Sequencing DNA was thought to be very difficult, if not impossible, until the late 1970s. However, at that time two different approaches were developed to sequence DNA. We will look at the more popular approaches here.

Procedure

First, we obtain a sample of DNA and make sure it is very pure. Then, the DNA strands are separated, and the sample of single-stranded DNA is split into four portions and placed in four tubes. DNA polymerase (the enzyme that manufactures new DNA) is added along with the necessary nucleoside triphosphates to each tube to allow replication to occur. If we were to let this mixture incubate, new DNA strands would be manufactured, which would be complementary to the original strands put in.

Here's the trick to sequencing. In addition to all four nucleoside triphosphates (one of which—generally dATP—is radioactive), we would add a dideoxynucleoside triphosphate (ddNTP) to each tube, a different one to each of the four tubes. The ddNTP is a nucleotide that can be used in DNA syntheses, just like a regular nucleotide (Fig. 6-1). But when it is put in the strand, no further synthesis of that strand can take place because dideoxynucleosides lack the 3′ –OH group, which is the "hook" on the sugar molecule that attaches to the next phosphate-sugar groups. So the synthesis of that strand stops right there. If we had added a dideoxy-adenosine triphosphate (ddATP), the synthesis of the strands in that tube would stop at the position A whenever that position was filled with dideoxy-A.

Because all the regular nucleoside triphosphates are present as well, synthesis continues on most strands. But some of the strands are shortened, having as their 3′ end the radioactive dideoxynucleotide. So in the tube with the dideoxyadenosines (ddAs), we have fragments of various lengths, each of them terminating with a ddA at their 3′ end. This is similar with ddT, ddC, and ddG in each of the other tubes. Figure 6-2 shows the result we will get in each tube.

The samples are then placed on an electrophoresis gel, each in their own lane, and electrophoresis is started. Each newly made

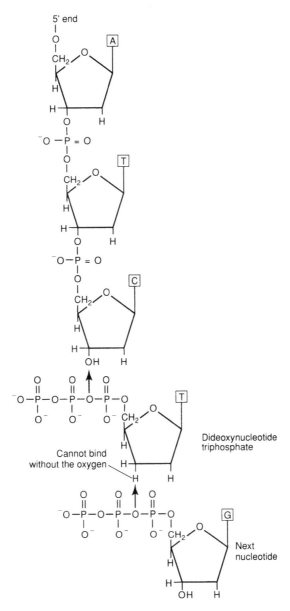

Figure 6-1 DNA synthesis in the presence of a dideoxynucleoside triphosphate (ddNTP). Ordinarily, synthesis of DNA occurs when a new nucleoside triphosphate is attached to the growing chain at the 3'-OH of the growing chain. When a ddNTP is added, the growth of the chain stops, because there is no oxygen (see note in figure) to which a new nucleotide can bond.

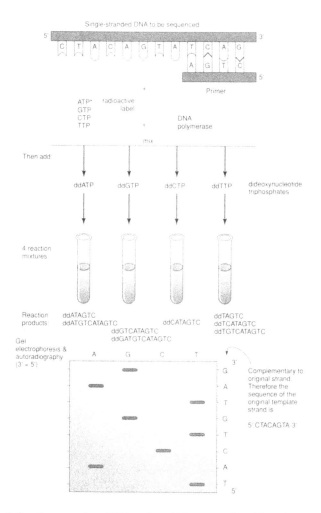

Figure 6-2 Sequencing DNA using dideoxynucleotides. At top is the length of DNA we wish to sequence by making a copy of it using DNA polymerase. (**a**) All the necessary nucleoside triphosphates (one being radioactively labeled) are added along with DNA polymerase. (**b**) In additional, a small amount of a single dideoxynucleoside triphosphate (ddNTP) is added—a different one for each tube. DNA is synthesized in each of the four tubes in accordance with the template, which is the DNA to be sequenced. (**c**) Fragments of the DNA are formed, since some of the synthesis is stopped because of the presence of the ddNTP. (**d**) The synthesis products (fragments) are put on a gel, and electrophoresis is performed to separate the fragments according to their length. Fragments differing in length by just one nucleotide can be separated. By looking at each lane and counting the number of nucleotides in the fragments in that lane, the position of each nucleotide in the unknown DNA can be determined.

DNA fragment moves through the gel, faster or slower, depending on their size. The result is a gel pattern that may look like the one shown in Figure 6-2.

The gel pattern can be "read" by looking at the lengths of the fragments in each lane. If conditions are right, a fragment will be in one of the lanes at each step, which in turn indicates which base is at that particular position in the DNA strand. In this way, it is possible to quickly learn the sequence of almost any piece of DNA. If the piece is too long to conveniently sequence, it may have to be fragmented initially and each of the fragments sequenced. DNAs containing thousands of nucleotides have been sequenced with this process.

For our purposes here, it is important to know the sequence of the DNA we are going to insert into the bacteria. Often, this DNA is one that contains the genetic message of the protein we want the bacteria to manufacture. We also want to make short DNA probes complementary to portions of this new DNA. So, sequences are important for us to have.

STARTING THE ENGINEERING PROCESS

To engineer the genes of bacteria, let's start by using bacteria containing a plasmid such as the one illustrated in Figure 5-1.

Engineering the Plasmid

The plasmid in the figure has already been sequenced to show the various sites where restriction endonucleases will cut (split) it, only some of which are shown. Note that two regions of the plasmid are marked Tc^r (tetracycline-resistant) and Ap^r (ampicillin-resistant). As noted previously, these are regions of the plasmid that confer resistance to the antibiotics, tetracycline and ampicillin. Resistance in both cases means that the bacteria will continue to grow in the presence of the antibiotic. For reasons that will become apparent as we proceed, it is important that we cut this plasmid using *Pst* I, which cuts the plasmid in the Ap^r region (Fig. 6-3).

We will use the opening in the Ap^r region to insert a new piece of DNA that we have obtained either by chemical synthesis or by cutting from another piece of DNA. The methods by which this can be accomplished have been outlined in Chapter 5. It is important that *Pst* I be used to cut the new DNA out of the original host or that the proper "sticky" ends have been synthesized

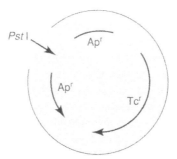

Figure 6-3   The plasmid shown in Figure 5-1, which has now been cut with *Pst* I. Note that by cutting it with this restriction enzyme, the ampicillin-resistant region (Ap^r) is broken, making the bacterium carrying the broken plasmid sensitive to ampicillin (ie, it will die if put in contact with the antibiotic ampicillin).

on the synthetic DNA, so that it will stick when placed with this cut plasmid (Fig. 6-4).

We then take the solution containing the cut plasmids and add the new DNA to the solution. We incubate this for a period of 1 to 4 hours at 16° C, so the pieces can find each other. The enzyme ligase is added, which will attach the new DNA insert to the plasmid DNA. The result is a plasmid that has a new insert in it. The insert is bonded to the DNA with phosphodiester bonds, which we discussed in Chapter 5. It is now an integral part of the plasmid.

Not all the plasmids will contain the insert, because some just combine with themselves, giving the original plasmid. Others may have multiple copies of the plasmid or multiple inserts. We need to segregate these various forms later.

Getting Plasmids into the Bacteria

The next step is to get the plasmid containing the insert into the bacteria. This can be done in several ways. We will mention two here. One method is to place bacteria in a medium containing **calcium sulfate**. When this is done, the pores in the cell membranes open up, allowing the plasmids to enter the bacteria. Calcium sulfate is one of several chemicals that will work to enlarge pores in certain strains of bacteria, giving plasmids relatively free entrance. Another method is to use **electroporation**.

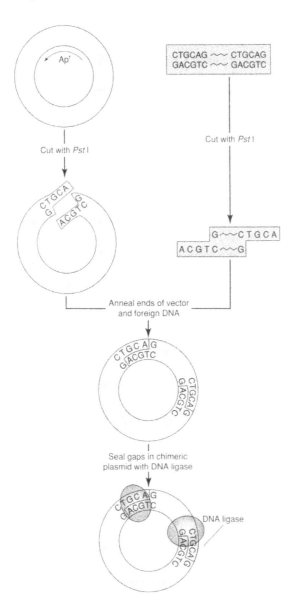

Figure 6-4 The use of *Pst* I to insert a fragment into a plasmid. The process is exactly the same as that described in Figure 5-13, except that *Pst* I is used. The reason we use *Pst* I is because this restriction enzyme cleaves the plasmid in the Ap^r region, which provides a way to identify which bacteria have the modified plasmids present. See text for details.

By subjecting cell membranes to a high-voltage electrical field for a short time, the membranes are temporarily broken down, producing pores large enough for plasmids to enter. After a period of time, the membranes realign and close the pores.

Using either method, plasmids can enter the bacteria. Once this is done, all we need to do is let the bacteria grow on a Petri dish and look for the bacteria that contain the new DNA.

SCREENING THE BACTERIAL COLONIES

How can we tell know which bacteria have the insert in their plasmids? An important trick allows us to easily learn this. Remember, plasmids came about in nature to give cells added features, such as resistance to antibiotics. We have chosen a plasmid that contains resistance to both ampicillin and tetracycline. So, with the plasmid containing both the Tc^r and the Ap^r regions, the bacteria would be able to grow on Petri dishes with nutrients containing both tetracycline and ampicillin. All bacteria that did not contain the plasmid insert would die.

If the plasmid inserted had been cut in the Ap^r region, as we wanted, it would disrupt the ampicillin resistance. The cell would then be susceptible to ampicillin, giving an Ap^s (ampicillin-susceptible) cell. Then, if we grew the bacteria on Petri dishes containing some ampicillin in the agar, we would kill all the cells that had our insert in them. This doesn't do us much good, because we would much rather kill all cells *except* those that contained the insert. So we need to do something different.

Replica Plating

Suppose that initially we grow the bacteria that we have transformed in a medium containing tetracycline. We will spread the bacteria on agar on Petri dishes so that individual bacteria are present at certain places on the agar. On incubation, these bacteria grow into colonies. When this is done correctly, each colony comes from a single bacterium, as shown in Figure 6-5.

Because the medium contains tetracycline, all bacteria that do not contain the plasmid will die, because only bacteria containing plasmids with a Tc^r region will live. So we know that all these colonies contain bacteria that contain the plasmid, but it could be either the original plasmid or one that we modified. What we want to do now is make several copies of the colonies on the Petri dish. To do this, we take a piece of velvet, cut it into

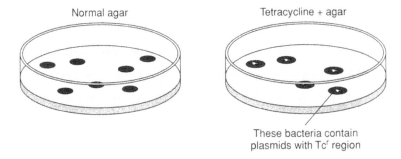

Normal agar   Tetracycline + agar

These bacteria contain
plasmids with Tc^r region

Figure 6-5   Drawing illustrating how antibiotics work. Bacteria grow on agar in Petri dishes in clusters called colonies. As the bacteria divide, they spread outward from the original bacterium in an almost circular fashion. If an antibiotic is present, the bacteria die, unless they contain a plasmid with the resistance gene to that antibiotic. In this illustration, tetracycline is present in the second Petri dish, and only those bacteria that contain a plasmid with the Tc^r region intact survive

a circle the size of the Petri dish, and then sterilize the velvet. By placing this sterile velvet on the original Petri dish, some of the bacteria stick to it and can be transferred directly to other Petri dishes. We can then grow the bacteria on these Petri dishes. These new Petri dishes will have the same pattern of colonies as the original. This is called **replica plating** (Fig. 6-6a).

Suppose we make two replica plates of the original, one on agar containing ampicillin and the other on agar containing tetracycline. By comparing the bacterial colonies, we can see those colonies that were not resistant to ampicillin and thus died. This tells us that the plasmids in those bacterial colonies contained the plasmid that contained an insert, since the insert destroyed the ampicillin resistance portion of the plasmid and made the bacteria susceptible to ampicillin.

Isolating the Strains

On the other replica plate, all the original colonies are grown, all of which have plasmids inserted and some of which are the modified plasmids. By taking samples of the bacterial colonies that died on the Petri dish containing ampicillin but lived on the tetracycline agar, we have isolated strains of bacteria that contain our insert! This approach to determining which bacteria contain the insert is called **screening** (Fig. 6-7), and the antibiotic-resistant sections of the plasmid are called **markers**.

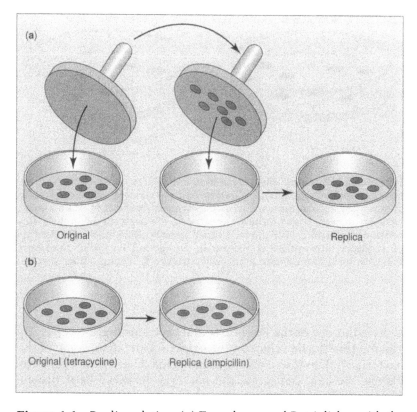

Figure 6-6   Replica plating. (**a**) To make several Petri dishes with the bacterial colonies exactly the same on each, a piece of sterile velvet (or similar material) is used to lift off a portion of the bacteria from each colony and transfer them to a new Petri dish in exactly the same location as they were found on the original Petri dish. In this manner, Petri dishes that have identical patterns of colony growth can be obtained. (**b**) Replica plating in which all the bacteria contain a plasmid (such as that shown in Fig. 5-1) that has both an ampicillin-resistant region (Ap^r) and a tetracycline-resistant region (Tc^r).

We have now placed engineered DNA into a living organism! When we put these bacteria in a culture flask, they will continue to grow and reproduce, replicating the plasmid DNA, which hopefully contains the DNA insert.

But there is still a question. We know that we have altered the plasmid in the Ap^r region, but we are not certain that the insert we want is really there. We need to perform some further tests to make sure the insert is really there.

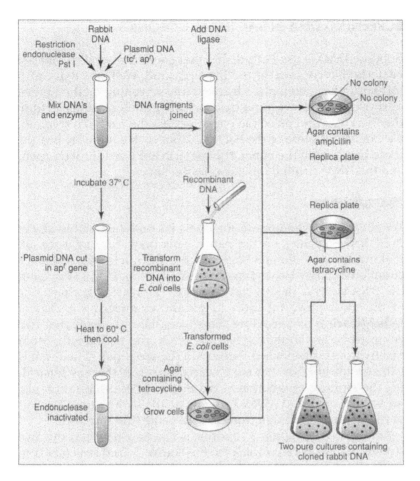

Figure 6-7 An illustration of the complete technique for inserting a piece of foreign DNA into bacteria and identifying (screening for) those bacteria that contain it. In this case, rabbit DNA and plasmid DNA are both cut with *Pst* I, incubated together, joined using DNA ligase and the plasmid solution mixed with bacterial (*Escherichia coli*) cells. To screen for bacteria that contain the plasmids with the rabbit DNA inserted, some of the bacteria are then transferred to Petri dishes that contain tetracycline. Only bacteria that contain plasmids will grow on these. Replica plates are made from the Petri dishes on agar containing ampicillin and tetracycline. The colonies that die on the ampicillin contain plasmids that have been cut in the Ap^r region, suggesting that these bacteria might contain the new DNA. These colonies are identified by their position and then removed from the other replica plate that contains tetracycline only. The bacteria are then cultured and grown. Additional screening, often by sequencing the plasmids, is needed to certify which bacterial colonies have the rabbit DNA in the correct position.

SCREENING DNA IN BACTERIAL COLONIES

Suppose that we use Petri dishes that contain several colonies of bacteria that contain the modified plasmid. We know that transformation has taken place by using the screening method previously outlined. If we cut a piece of nitrocellulose paper to fit in the Petri dish and place the paper against the bacterial colonies, we can lift off some of the bacteria from each colony. By heating these bacteria on the paper, the bacteria are lysed (broken open) and the DNA within them sticks to the paper.

Hybridization

We now want to find out if our insert is contained within any of the DNA molecules on the paper. This may seem as if we are looking for a needle in a haystack, but the nucleotide sequence complementarity (ie, the fact that A sticks to T, and G to C) of the DNA strands themselves comes to the rescue. Perhaps one of the most powerful tools in genetic engineering is that of **hybridization** or annealing. This term merely means that two complementary strands of nucleic acids tend to find each other and hydrogen-bond when they are in the same vial. Whether it is two strands of DNA that have been denatured, thereby separating the strands, or whether it is a piece of DNA that is complementary to a strand of RNA, the principle is the same. For our purposes, we can design experiments using this approach.

As noted in Chapter 3, hydrogen-bonding between the two strands of DNA is responsible for the double-helical structure G to C, A to T). Although the hydrogen bonds (H-bonds) are relatively weak, numerous H-bonds provide substantial bonding between two strands of nucleic acids. So the design of the screening techniques that we will outline in the following section is based on the principle of bonding or hybridizing two strands of nucleic acids together because of their complementarity.

Making a Probe

We first make a short piece of DNA that is complementary to a portion of the DNA that we have inserted into the plasmid (Fig. 6-8). This short piece of complementary DNA is called a **probe** and should be able to hybridize to the portion of any DNA to which it is complementary, provided that the strand of DNA is available for hybridization. Heating the double-helical DNA on the nitrocellulose paper breaks the H-bonds between strands and opens up the DNA in a proper fashion. To be useful in this

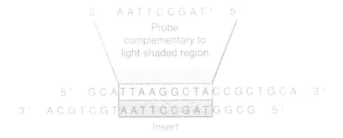

Figure 6-8. Hybridization. Whenever a new piece of DNA is hydrogen-bonded to a complementary piece of DNA, the product is called a hybrid. Short pieces of DNA can be synthesized chemically with any desired sequence of bases in them. These new, short pieces can be added to existing DNA strands and hybridize with them in those locations in which they are complementary, as shown here. If the existing DNA is in double-stranded form, the new DNA can still hybridize with it, forming a triple strand. Often the short DNA probes are made radioactive to allow identification.

search, the probe must be labeled, often with a radioactive label such as ^{32}P (radioactive phosphorus). This will allow us to find the probe when we use it to search for DNA.

The minimum probe length needed for hybridization is about 10 nucleotides, but the size can vary, depending on hybridization conditions. The longer the probe, the more stable the hybridization. Therefore, longer probes are generally used to ensure that hybridization will take place. There is a limit to this, because very long probes can interfere with hybridization or part of them can hybridize to the wrong regions of the target DNA. To ensure that the hybridization is unique for the gene for which we are looking, probes of 20 to 25 nucleotides are generally used.

The nitrocellulose paper containing the lysed colonies of bacteria is then put in a plastic bag, and a solution containing the radioactively labeled DNA probe is also put in the bag. This probe has a sequence complementary to the sequence of the new DNA insert in the plasmid.) The bag is sealed, and the solution is swirled around. The nitrocellulose paper is then removed, and the solution is washed off to remove all probes that aren't stuck to complementary DNA. The paper is dried and then placed on top of a piece of x-ray film and put in a light-tight container and allowed to stay there for a few hours (Fig. 6-9).

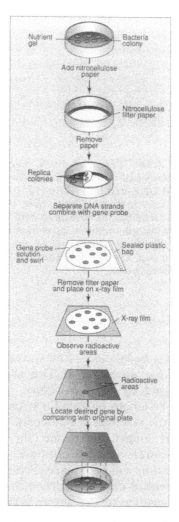

Figure 6-9   The colony screening technique. This method is designed
to identify bacteria that contain the new DNA (see Fig. 6-8). After the
bacterial colonies are grown, nitrocellulose paper is used to lift off
some bacteria from each of the colonies, giving a replica of the pattern
of colonies on the dish. The nitrocellulose paper is heated to lyse the
bacteria and separate the DNA strands. It is then put in a sealed plastic
bag along with a solution containing a short piece of radioactively
labeled DNA, which is complementary to some portion of the inserted
DNA. After mixing for a short time, the nitrocellulose paper is removed
and washed and then placed on x-ray film. The film is exposed wherever
the radioactive DNA probe has hybridized with the new DNA in the
bacteria. In this way, the colonies of bacteria containing the new DNA
are screened and identified.

The radioactive DNA probes will expose the x-ray film, causing a dark spot on the film wherever they are bound to the DNA of the plasmid insert. Thus, the pattern of spots on the x-ray film will reveal which, if any, colonies of bacteria contained the new insert (Fig. 6-9). This is called the **colony screening technique**.

After the colonies that have the correct insert are identified, these bacteria can be obtained from the original Petri dish and put in culture and substantial numbers of these engineered bacteria grown. Using some of these bacteria, the plasmids could be isolated directly and the sequence of the plasmid determined to certify the absolute presence of the correct insert.

Screening Specific DNA Fragments (Southern Blotting)

Another way to identify a specific gene in a piece of DNA is to screen the DNA particles directly. Each of the different bacterial colonies can be isolated, grown up, and lysed, and the DNA in the bacteria extracted. This DNA is then cut, using one or more restriction endonucleases. The fragments are placed in an electrophoresis gel apparatus and separated according to size, as outlined in Chapter 5.

Because DNA in the gel itself is not really available for hybridization with a probe, a piece of nitrocellulose paper is placed on the gel and the DNA is "blotted" out of the gel onto the paper.

The paper is then heated to open the double-stranded DNA fragments, after which it is inserted into a plastic bag and swirled with a solution of radioactively labeled DNA probes. These probes are complementary to the piece of DNA that was inserted into the plasmid. Upon washing the paper and exposing the x-ray film, the radioactive label will expose the x-ray film at those sites where it binds to the plasmid fragments containing the new inserts. This technique is called **Southern blotting**, named after its inventor, Dr. Edward Southern (Fig. 6-10).

PRODUCING PROTEINS FROM GENETICALLY ENGINEERED BACTERIA

One of the major purposes of engineering bacteria is to make large amounts of a specific protein. Identical bacteria that are replicating without conjugation or any other means of altering their genetic information are **clones**. So whenever we take some

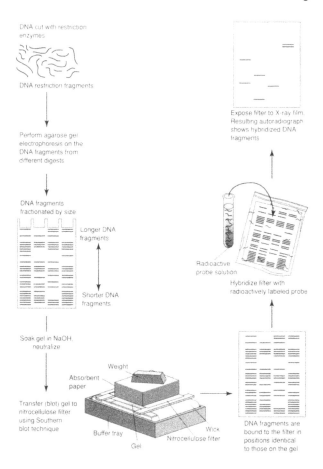

Figure 6-10 The Southern blotting technique. This technique is very useful to identify fragments of DNA containing a known sequence of nucleotides. The DNA is initially extracted from a cell and then digested using one or more restriction enzymes. Gel electrophoresis spreads the fragments according to size. The DNA fragments are "pulled" from the gel by placing the gel on a platform with a piece of nitrocellulose filter paper above it. Absorbent paper is placed above the nitrocellulose paper to "pull" the solution and DNA through the gel and to the nitro-cellulose paper. The result is a pattern of the bands from the elec-trophoresis on the nitrocellulose paper. This is now heated and swirled in a bag with the radioactively labeled DNA probe, which is comple-mentary to regions of the DNA fragments. The fragments containing the right sequence can be identified by placing the nitrocellulose paper on a piece of x-ray film and letting the radioactivity expose the film in the bands containing the probe.

bacteria and spread them around on a Petri dish until there are only single cells left, the colonies that grow from that single cell are clones of the original bacterium. If there is a plasmid with an insert in the original bacterium, we then have clones of a single, engineered bacterium.

Suppose that the DNA we inserted into the plasmid is really the gene for a specific protein, such as insulin. Insulin is used to treat diabetes, so lots of insulin is used throughout the world. Although insulin from other sources, such as pigs, has been used to treat diabetes, it doesn't always work well. Human insulin is highly desirable. So, suppose that the human insulin gene is inserted into a plasmid and the plasmid is put into bacteria that are cloned. When that gene is translated into protein (**expressed**), the insulin we wanted will be made by the bacteria, as well as the usual bacterial proteins. Multiple copies of the gene may be inserted as well, giving what is called a **high copy number** plasmid. This allows numerous identical proteins to be made by the bacteria. They literally become "protein factories." So, we can harness bacteria to do our work for us.

In principle, we should be able to take any piece of DNA within reasonable size limits and put it into bacteria. Therefore, if we were to take a gene that makes any protein we want, we could place it in a plasmid and insert that plasmid into bacteria. Thus, we should be able to make enormous amounts of that particular protein. Unfortunately, it doesn't always work that way.

There are some proteins, especially those used by humans, that bacteria do not make well. So we need to find other ways to make this process work. We will discuss this more in Chapter 8.

MAKING GENE LIBRARIES

There is one final issue to discuss. We have mentioned that we may need to obtain a gene from natural sources, but how do we go about doing this? At first, it seems a formidable task. For instance, in the human genome there are 23 chromosomes, and each contains millions of nucleotides. How are we to find a gene for a certain protein? Again, let's use the protein hormone insulin as an example. This protein is made specifically by pancreas cells. Somewhere, in all the DNA contained in the pancreas cells, the insulin gene is buried. How do we proceed?

Library of Genes

What we really need to do is make available all the DNA in the pancreas cells in a manner that we could sort through it with a probe, looking for the insulin gene. In essence, we need to create a library of bacteria, each one containing a small portion of the pancreas gene. To do this, we first get a pancreas cell line and then grow a large number of these cells in tissue culture. (We will discuss this technique in Chapter 8.) After that, we disrupt the cells and isolate their DNA. Using a set of restriction enzymes, we then cleave the DNA into fragments of sizes that could be inserted into plasmids. Plasmids obtained from bacteria are then cut using the same set of enzymes, following which the plasmids and the restriction fragments of the pancreas cell DNA are incubated together.

The fragments of pancreas DNA will anneal to the sticky ends of the plasmid DNA and can be ligated in place. Most of the pancreas DNA fragments will be incorporated into bacterial plasmids. These **chimeric plasmids** (plasmids with inserts in them) are then transferred into bacteria, and the culture is grown. The result will be that numerous bacteria with a great variety of plasmids will be in the culture (Fig. 6-11).

Ideally, *all* the DNA fragments from the pancreas cells should be found among all these bacteria. These bacteria are called a **gene library** of the cell chromosome because they house all the genetic information found in that cell line. Just like a library filled with books, the individual bacteria contain certain fragments of the pancreas cell genetic information. But do we find the insulin gene?

First, we grow colonies of bacteria containing the various plasmids that have inserts from the pancreas cell DNA. These colonies can then be screened for the particular gene of interest, using the techniques described previously. We would make a probe complementary to the insulin gene and then use colony screening techniques (outlined earlier in this chapter) to find the bacteria that contain the insulin gene. Then, we take the bacteria containing the insulin gene, grow them, isolate their plasmids, and then purify the insulin gene from the plasmids. This gene can then be inserted into other plasmids, reinserted into bacteria, and will then be expressed in those bacteria giving— human insulin.

Gene libraries from many cell lines are now available, making a remarkable resource. As the human genome is sequenced, many additional libraries are being made. This develops a

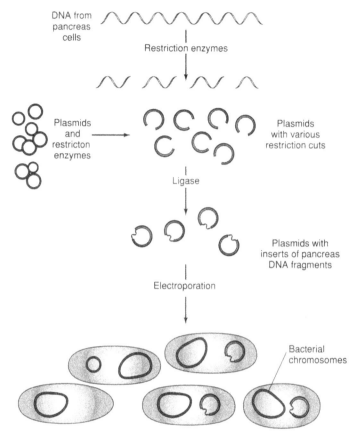

DNA from
pancreas
cells

Restriction enzymes

Plasmids
and
restricton
enzymes

Plasmids
with various
restriction cuts

Ligase

Plasmids with
inserts of pancreas
DNA fragments

Electroporation

Bacterial
chromosomes

Bacteria can contain plasmids with inserts,
plasmids with no inserts, or no plasmids at all

Figure 6-11 Making a gene library. The purpose is to make a family of bacteria that each contain a single fragment of DNA from a given source and together contain all the fragments from the DNA source. In the example, the complete DNA from pancreas cells is cut up using a mixture of restriction enzymes. This mixture of restriction enzymes is also used to cut plasmids that have been isolated from bacteria. The DNA fragments from the pancreas are then mixed with the cut plasmids and DNA ligase is added to complete the process. The result is a large number of plasmids containing various pieces of the pancreas DNA. These plasmids are then inserted into bacteria by electroporation, a technique that electrically opens pores in the bacteria. The bacteria that result will contain the plasmids that in turn contain all the fragments of pancreas DNA inserts. Then, when a certain portion of pancreas DNA is needed, the bacteria can be put on Petri dishes and the colony screening technique outlined in Figure 6-9 can be used to identify the bacteria containing the portion of DNA needed.

tremendous gene bank to be called on as needed. We will further discuss how these genes can be used in human genes cells in Chapter 12.

APPLICATIONS

Several genetically engineered proteins, notably insulin, interferon, and tissue plasminogen activator (TPA), are being made by bacteria and are now sold in large quantities. The quality and quantity of these engineered proteins are high and the cost is low, compared with earlier costs when they had to be extracted from tissue. For instance, the cost of interferon was enormous— about $10 million per gram before it was cloned. Now the cost is but a few hundred dollars per gram.

Harnessing bacteria to produce proteins important to humans can be relatively inexpensive and extremely useful. However, there are limitations, because genes of mammals contain extra pieces of DNA (introns) and other alterations, which bacteria may not be able to translate. In these cases, it has become necessary to use mammalian tissue culture to make the desired proteins. This will be discussed more fully in Chapter 12. Nonetheless, the approach in which desired genes are cloned into bacteria will continue to be an important application for genetic engineering techniques (Table 6-1).

Table 6.1 Genetically Engineered Pharmaceutical Products

Product	Originator(s)	Sales ($ Billions)	
		United States	World
Erythropoietin	Amgen, Genetics Institute	600	1125
Hepatitis vaccine	Biogen	260	724
Human insulin	Genentech	245	625
Human growth hormone	Genentech, Biotechnology General	270	575
Alpha interferon	Genentech, Biogen, Wellcome	135	565
Granulocyte colony-stimulating factor	Amgen	295	544
Tissue plasminogen activator	Genentech	180	230
Granulocyte-macrophage colony-stimulating factor	Immunex, Genetics Institute	50	70
Gamma interferon	Genentech, Biogen	15	25
Interleukin-2	Immunex, Chiron	5	20
Total		2055	4503

SUMMARY

Genetic engineering of bacteria means that we insert new DNA of a known kind into bacteria after first sequencing the DNA. The method allows us to read the sequence directly from an electrophoresis gel pattern. We can then insert the new DNA directly into a plasmid, using the right enzymes to cut open the plasmid and seal the new DNA in place. The new DNA is placed in a region of the plasmid that provides antibiotic resistance to the bacteria containing the plasmid, and the modified plasmid is put into the bacteria. The antibiotic resistance disappears, allowing us to know which bacteria contain plasmids with new DNA.

Various methods are available to screen (isolate the bacteria with the right DNA) the bacteria. Colony screening allows us to grow colonies of bacteria and identify those that have the new DNA present. Southern blotting is also a process by which we can extract the DNA from the bacterial colonies, fragment it, and test it to find out whether the new DNA is present.

When the new DNA is in the bacteria, we can grow enormous quantities of these bacteria, all of which contain the new DNA. These clones not only make new bacteria like themselves, but also manufacture proteins for their use. By adding numerous repeats of the new DNA, the bacteria then makes lots of new protein, which could be needed by man (such as insulin).

Gene libraries are made to find the genes that will make proteins useful to man, which can contain most of the genes from specific human cells, such as the pancreas cells. Once located, the genes can be prepared and purified to allow them to be inserted into plasmids and used to engineer bacteria.

7

VIRUSES

WHAT YOU WILL LEARN IN THIS CHAPTER

- How specific viruses attack specific cells
- How bacteriophage (phage) reproduce themselves
- How active phage can be reconstituted from phage protein and phage DNA
- How cosmids can be used to bring new DNA into cells
- How to detect the presence of the new DNA in phage by means of screening techniques

---

Viruses attack all living things and often cause disease or even death. Viruses are really a stripped-down version of a living thing. They are not really alive, at least according to the definition we applied in Chapter 1. Viruses have some of the tools needed to carry out functions of living things, but they depend entirely on other organisms to give them the power to do so. They are not able to make it on their own.

A bacterial virus is called a **bacteriophage** (bacteria-eater) or, more often, **phage**. Some viruses have more components, but the simplest have just a **nucleic acid genome** and a protein covering. A schematic picture of a bacterial virus is shown in Figure 7-1. You can see DNA enclosed by a wall of protein.

Viruses are generally tailored to use a particular target cell as a **host** (an organism in which they can reproduce themselves). There are various categories of viruses—some much more complex than others. The simplest viruses, like the one in Figure 7-1, use bacteria as their hosts. Their sole purpose in life is to find a particular bacterium, attach to it, insert their DNA, and use the bacterial mechanisms to reproduce more phage. In this

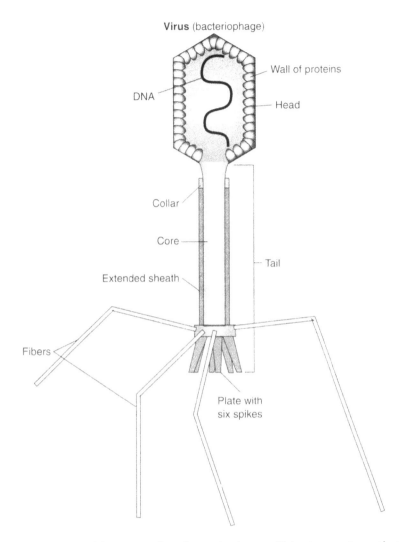

Figure 7-1 Diagram of a bacteriophage. This is a virus that specifically attacks bacteria. The phage seems to work like a syringe with which to insert the nucleic acid genome of the phage into the bacterium.

way, the phage doesn't have to carry around all the excess machinery needed to reproduce itself. It borrows the machinery of the host cell.

Bacterial viruses are very important in genetic engineering because they contain genetic material and because they know how to get it into the host cells. This makes them ideal trans-porters of genetic information—even genetic information that has been placed there by us. The advantage of a virus over a plasmid is that the virus can carry larger genes than can the plasmids, so they are more suitable for some tasks. In addition, viral DNA is generally placed directly into the host chromo-somes, often in active regions. For this reason, we need to look a little closer at how a bacteriophage works.

THE LIFE CYCLE OF A PHAGE

Initially, a phage attaches itself to its particular host bacterium, then inserts its DNA into the host in much the same way that a hypodermic syringe and needle are used to inject fluid into us. The phage DNA can do one of two things within the bacterium, depending on the kind of host the bacterium is.

The Lytic Pathway

In the first case, the phage DNA may just go about its business of getting its genetic message out and getting new viruses made. The viral DNA is also reproduced in multiple copies to make new DNA for the new virus particles. When new viruses are to be made, it uses the cell's machinery to make the viral proteins necessary for the new virus particles. The genetic information used to make these proteins is encoded in the DNA that the virus put into the cell's chromosome.

All the cell's energy is used to make new phage particles, DNA and all. The result is that, after a little while, the phage has made many copies of itself, normally from 100 to 200, and the cell walls break open, releasing the new phage. These new phage then go about looking for new cells to infect. The bacteria are in trouble!

This process is called the **lytic pathway** of a bacteriophage and is diagrammed in Figure 7-2. In this case, the DNA of the phage is used directly to make more phage. This reproduction of the phage causes massive destruction of the bacteria in a very short period of time (hours), since each infected bacterium releases from 100 to 200 viral progeny as it is lysed (broken open).

In higher-order cells and organisms, similar processes occur when they are attacked by a virus, but reproduction takes much longer (days). Sometimes the host cells are not destroyed, but produce new viruses over a period of time and send them out one or a few at a time.

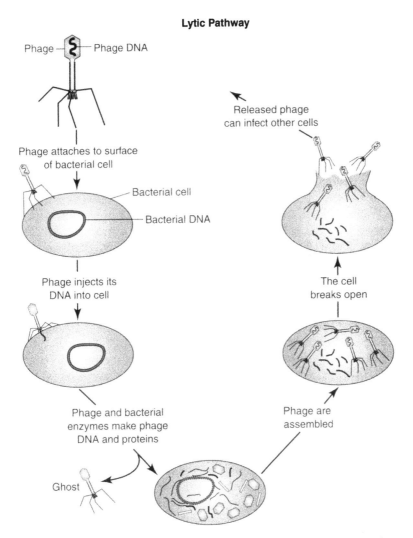

Lytic Pathway

Phage — Phage DNA

Phage attaches to surface of bacterial cell

Bacterial cell

Bacterial DNA

Phage injects its DNA into cell

Phage and bacterial enzymes make phage DNA and proteins

Ghost

Phage are assembled

The cell breaks open

Released phage can infect other cells

Figure 7-2 The lytic pathway. New bacteriophage are produced as diagrammed, each of which is identical with the original phage. In a typical lytic cycle, about 100 to 200 new bacteriophage are produced.

The Lysogenic Pathway

A second, more subtle, process by which phage can be produced is called the lysogenic pathway. In this case, the DNA of the phage is combined with that of the bacterial chromosome as before. The viral DNA is placed in the bacterial chromosome without the bacteria knowing it. This hide-and-seek game is very effective. Sometimes the viral DNA will stay in the bacterial chromosome for many bacterial generations. The viral DNA is replicated along with the bacterial DNA in each generation until one day the following happens: Triggered by some **inducing function** (e.g., UV light) that stresses the bacterium, the viral DNA is activated and starts to make proteins and also more viral DNA—and the phage infection is on its way. This process is called the **lysogenic pathway** (Fig. 7-3). It is this process that is most useful to us in genetic engineering, because it provides a way to insert new DNA into the chromosome of a bacterium in a rather permanent fashion.

We should mention here that the virus is not always accurate in the lysogenic process. Sometimes in the natural process of getting its viral DNA out of the cell, it takes some cellular DNA with it. Then, when it infects the next cell, it carries information from the previous host with it. The new cell receives new bacterial DNA, which may give it additional mechanisms to adapt to new environmental stresses. This is another important way by which bacteria are able to adapt to the environment around them. The process is called **transduction**.

Although the lysogenic mechanism can spell doom for bacteria, it is very useful for genetic engineering, as we will shortly see. If we can somehow get the virus to carry in some extra DNA and insert this with its own DNA into the bacterial chromosome, then we can get a piece of new DNA into the DNA of the cell and let the cell do all the work of replicating the DNA. This is at the heart of what we wish to do in genetic engineering.

It should be emphasized that only certain bacterial strains can become lysogenic, so the bacteria play an important role in the process. When a bacterium has received the phage DNA into its chromosome, it is called a *lysogenic bacterium*.

There are a couple of ways in which viral DNA can be used to carry engineered pieces into the bacteria. In earlier years, scientists found that DNA could be removed and the virus altered and then reinserted. In more recent times, another method has been developed in which the coat is really not used. We will discuss each method separately.

PHAGE RECONSTITUTION APPROACH

The first thing we need to do is extract the viral DNA and get it where we can use it. This is easy because viruses like to get rid

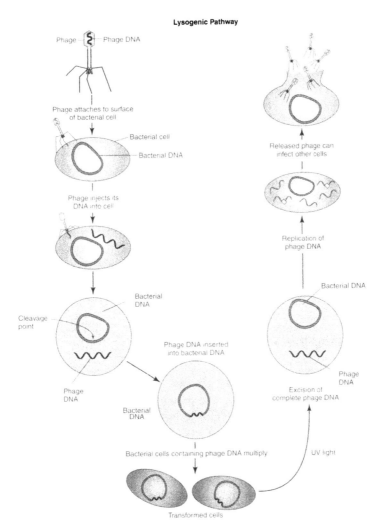

Lysogenic Pathway

Figure 7-3 The lysogenic pathway. In this process, the DNA inserted by the bacteriophage is actually placed in the chromosome of the bacteria and stays there for an undefined period of time. Cells containing phage DNA are called transformed cells. If these cells are induced by some event (ultraviolet light does the trick), the phage DNA is excised from the bacterial DNA and the process of making new phage particles begins, as it did in the lytic process. The only difference is that in this case, there are sometimes portions of the bacterial DNA that get connected with the phage DNA, so the phage can actually carry bacterial DNA from one bacterium to another. For our purposes, the important process is the ability of the phage DNA to insert its DNA into the bacterial chromosome.

of their DNA anyway and also because the DNA is not bound tightly in the virus. The virus particles are treated with an enzyme that breaks down the protein coat, and the DNA is then released into the solution.

Many experiments today use a well-characterized phage, the **lambda phage** as the carrier, because its DNA is fully sequenced. So, once the protein coat is removed and we have selected a site in the lambda phage DNA in which to insert our piece of DNA, we use a restriction endonuclease (enzyme) to cut open the phage DNA at that site. As with the plasmid approach, we must be certain that the new DNA to be inserted has ends similar to those on the cleaved phage DNA. We then put our new DNA and the endonuclease-opened phage DNA together in a tube and bond them to each other, as outlined for plasmids in Chapter 6. Of course, not all the DNA particles will recombine the way we want them to, so we will have to screen the products eventually as we did before. But for now, let's assume that the DNA is put in where we thought it should be.

Rebuilding the Virus

We now need to rebuild the virus. This is not as hard as it sounds. We can obtain **viral ghosts**—viral coats without DNA in them—by gently breaking the viruses open and removing the DNA. These ghosts (which have no DNA inside) are then put with the new viral DNA, and new viruses are reformed spontaneously. It is even possible to extract just the viral coat proteins from intact viruses and put these together with the DNA and form new virus particles. Although not all the viruses reform, some do, and these are then able to be used to insert our DNA into susceptible bacteria.

There is one problem with this approach. The phage heads have a limited amount of space available, and most of that space is taken up with the phage DNA. So the lambda phage has been pared down by the removal of nonessential genes to make room for the inserts. In addition, the lambda phage contains a single EcoRI site, where an insert can be made. The phage DNA has now been engineered to the degree that unless it contains an insert, the DNA will not go back into the phage head. So only those pieces of DNA that contain the insert (**chimeric DNA**) are reunited with a phage head giving virus particles (Fig. 7-4). Thus, DNA that is too short or too long will not be accepted by the phage ghosts. This provides a nice screening method, since

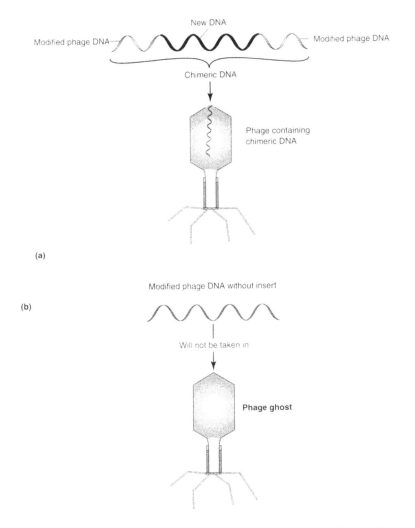

Figure 7-4 Insertion of a chimeric DNA into a bacteriophage. Since bacteriophage can hold only a certain amount of DNA. (**a**) By modifying the phage DNA and inserting new DNA of the proper length, the resulting chimeric DNA can be readily inserted into the phage ghost. (**b**) But if the DNA insert is not present, the DNA will not be taken in by the phage ghost. A phage ghost is the emepty phage (no DNA).

only DNA pieces of the correct size are able to form viable viruses, and these are the DNA pieces that contain the inserts.

The rebuilt viruses are then used to infect lysogenic bacteria so that the new gene is inserted into the bacterial chromosome.

The bacteria remain in the lysogenic state, carrying the new DNA with them. The important question is whether or not the inserted gene will be **expressed** into a useful protein. As noted before, genes are expressed when there is a promoter region (the region that starts the messenger ribonucleic acid [mRNA] transcription) in front of them. If we are careful about where the phage DNA is inserted into the plasmid or the bacterial chromosome, we can then get our DNA to be expressed a little or a lot.

One of the ways to make sure the gene will be turned on is to include a promoter region with the gene that is to be inserted. Then, by "turning on" the promoter region, the gene is transcribed into mRNA, which is made into a protein. It is also possible to insert many genes behind a promoter region. In some cases, over 1000 copies of the genes of a single protein gene have been inserted into the bacterial chromosome. The bacterium then becomes a little factory for that kind of protein, allowing considerable quantities to be made inexpensively.

COSMID APPROACH

Although a phage will allow more new DNA to be inserted than occurs with a plasmid, the phage is sometimes still not large enough to accommodate large or multiple-copy genes. To allow even more DNA to be inserted into bacteria, the cosmid approach has been developed. A **cosmid** is a hybrid (cross) between a plasmid and viral DNA. It is really a plasmid into which has been inserted the DNA sequences needed to package the plasmid into a virus. These viral DNA sequences are called *cos* **sites**. The remainder of the plasmid can carry the new DNA and any other information desired, but is still limited in size by the quantity the phage can carry. However, more information can be put into a cosmid, since very few phage genes are needed. These chimeric DNAs (those containing the insert) can then be purified by packaging them into phage heads. And they can then be used to infect bacteria. Because the cosmids are primarily plasmids, they enter the bacteria and form plasmid-like structures within the bacteria, which ordinarily do not enter the chromosome. They can be thus treated exactly as we have treated the plasmids previously. The advantage is that cosmids can carry more information than the phage inserts normally do.

Once inserted into the bacteria, the cosmids are perpetuated as plasmids. Yet they can ultimately be isolated and packaged back into phage particles by merely adding the coat proteins of the phage, using a helper phage to insert the genes for the necessary protein into the bacteria. These new phage are extremely useful as vectors (carriers) to bring new DNA into bacteria.

SCREENING FOR VIRAL INSERTS

How do we find out which viruses have the correct inserts? As already noted, in some cases we can make the selection based on size alone, because only those genomes with the inserts will fit in the phage head. In other cases, we may need to screen for the inserts.

To do this, the phage are placed at specific, marked locations on duplicate Petri dishes on which a "lawn" of bacteria has been grown. When this is done, the phage grow on the bacteria, leaving holes in the lawn called **viral plaques**. These are regions in which the bacteria are dead, but the viruses are numerous.

Each plaque contains phage, but some of the phage particles do not contain the correct DNA insert. How do we tell them apart? If we place a nitrocellulose paper on the agar in one of the replica Petri dishes, the phage stick to it. Then, we can wash the paper in an alkaline solution, which will lyse (break open) the phage and denature (break the hydrogen-bonds, etc) DNA that was in the phage. The DNA remains stuck to the nitrocellulose filter paper. This entire process is shown in Figure 7-5.

To identify plaques that contain the new DNA, we perform the same steps as for the bacterial screening method: We put the nitrocellulose paper in a plastic bag, swirl our labeled probe solution with the paper, wash the paper, and use it to expose an x-ray film. The plaque or plaques containing the new DNA will show up as dark spots on the x-ray film.

OTHER VIRUSES

There are numerous viruses known, many of which attack primarily eukaryotic (higher, more complex forms) cells. Each kind of virus replicates itself in specific ways. Many cause sickness and even death, since viral infection is often difficult to control once it starts. For example, common colds are caused by viruses. Influenza is viral-induced. The AIDS epidemic is caused by the human immunodeficiency virus (HIV). And all around the world, new viruses such as the Ebola virus, which caused massive death in Africa, seem to be cropping up. And none of us

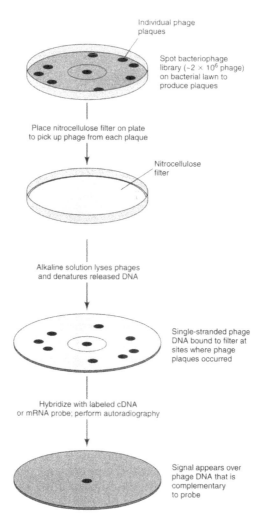

Figure 7-5 Phage screening technique. First, a Petri dish is inoculated with bacteria all over it and allowed to grow, giving a "lawn" of bacteria. Phage that have received DNA inserts are then spotted at various locations on the bacterial lawn. The phage are allowed to grow, killing the bacteria and producing plaques (places where the bacteria are killed by the phage infection). The viral plaques are lifted off by a nitrocellulose filter paper, after which the paper is treated with an alkaline solution that lyses the phage, releasing the DNA (which sticks to the paper in those locations). The paper is then swirled with a radioactively labeled probe complementary to a portion of the DNA insert, and those plaques containing the insert are identified. The phage containing the insert can then be removed from the original Petri dish and used to infect more bacteria, producing enormous amounts of phage.

is unaware of the enormous problem developing because of the HIV virus.

Although many viruses are pathogenic, many (including HIV) have also been harnessed as tools in the genetic engineering effort. We will discuss this in more detail in the next chapter.

SUMMARY

In this and the previous chapter, we have discussed two methods of bringing new DNA into bacteria—by using plasmids and by using bacteriophage. Phage have some advantage, because they can bring in more new DNA than plasmids can. In addition, phage can have a life cycle (lysogenic) that allows them to insert their DNA into the bacterial chromosomes, where it is left for many generations. This allows us to put new DNA into the phage and let the phage insert it into the bacterial chromosome. The result is that bacteria then contain the new DNA, which, when it is expressed, makes proteins that are specified by the new DNA. This approach gives us an additional way (along with plasmids containing new DNA) of harnessing bacteria to manufacture proteins of use to mankind.

Some proteins can be made using only cells from higher organisms. Using the same approach outlined in this chapter, we can use other viruses, which have been modified to contain new DNA, to insert that DNA into higher-order cells.

III

MAKING GENETIC CHANGES IN PLANTS AND ANIMALS

8

PLACING NEW
GENES IN
MAMMALIAN CELLS

WHAT YOU WILL LEARN IN THIS CHAPTER

- How transfection (putting new DNA into mammalian cells) is carried out
- How to screen mammalian cells to ensure that the new DNA is present
- How new DNA can be inserted into mammalian cells using viruses as carriers (vectors)
- How to make complementary DNA (cDNA) libraries of certain cell lines

---

One of the nice things about bacteria is that they are easy to manipulate and quick to grow. This in turn allows genetic trans-formation to occur easily. The growing process is much more difficult with plants and animals than with bacteria, because generation time increases and the complexities are greater. However, in recent years, it has been possible to extract certain kinds of cells from animals, including humans. Through careful manipulation, these cells can grow and divide in culture. This process is called **tissue culture**. When cells are cultured in this way, they act a lot like bacteria and can be manipulated using the tools we have previously described in Chapter 6 and 7, at least to some degree. This has been a tremendous benefit to those wanting to work with the higher-order (eukaryotic) cells. One major difference between a bacterial cell and a eukaryotic cell is that very few plasmids (see Chapter 6) exist in eukaryotes, so the process of altering the DNA happens mainly in the chromosome itself.

Why do we want to go to all the trouble of manipulating eukaryotic cells, when bacteria are so much easier to manipulate? The answer is that bacteria, as good as they are, cannot be induced to express many mammalian genes. Some mammalian proteins are substantially more complex than those found in bacteria. Because of this, it is often necessary to harness eukaryotic cells to do the work. In addition, some eukaryotic proteins have additional sugar units attached after proteins are manufactured, which bacterial cells cannot do.

In the final analysis, we want to learn to manipulate human cells, because we would especially like to engineer them in order to produce proteins that would be useful to mankind. An example of this is the manufacture of tissue plasminogen activator (TPA), a protein that will dissolve blood clots in patients who develop them. This protein was the first to be made through genetic engineering of animal cells. Now, rather than having to isolate TPA through laborious and expensive extraction processes from an animal source, cells containing the gene insert for this protein can be grown in tissue culture, and massive amounts of the protein can be isolated and used. We will more fully discuss this and other examples later in this chapter.

TRANSFECTION

It has been known for years that DNA in tissue culture medium can be taken up by the cells of mammals (this includes humans). The process by which this is done and incorporated into the genome of the cells is called **transfection**, the eukaryotic counterpart to transformation.

Calcium Phosphate Method

One way transfection happens is that DNA is carried into the cells with a bit of calcium phosphate. Similar to what happens in bacterial transformation, the technique is to prepare a transfection mixture of the DNA sample and then add calcium phosphate. Special techniques have been developed that allow up to 20% of the cells to be transfected.

Electroporation Method

Probably the most common method of transfection is to use electroporation (see Chapter 6). As with bacteria, electroporation disrupts the cell membrane and opens pores that allow DNA to enter. DNA pieces up to 150 kb long can be inserted in this way, with a minimum amount of preparation. The efficiency of this process is about 10% at most. In both this and the calcium phosphate approach, the DNA is inserted randomly into the cell. So a major question is whether the DNA is actually incorporated into the gene, and, if so, is it incorporated in a place where it can be used. Because this method is so easy to use, it is a very popular way to transfect mammalian cells.

MARKING AND SCREENING TECHNIQUES

Generating Thymidine-Deficient Cells

As noted in Chapter 6, one of the challenges in genetic engineering any cell us to find out when the new DNA insert is actually in the cell. With bacteria, we were able to use the convenient antibiotic markers on the plasmids. However, mammalian cells don't have plasmids, so the question we must answer is when we have put new DNA into the chromosomes. Close on the heels of this question is whether we have put it in a place where it will be used to make proteins, which means it should be put behind a promoter region. To screen for the presence and the placement of inserts, it is necessary to develop some rather tricky ways to screen mammalian cells.

The problem of random insertion has plagued researchers for a long time, and they have expended considerable effort to develop suitable markers and screening techniques. There are now several ways in which the presence of the insert into the DNA of the cell can be screened. One of the most common is using the **thymidine kinase** (tk) gene as a **marker**. Thymidine kinase is an enzyme that adds a phosphate group to thymidine (T is one of the four nucleotides), making thymidine monophosphate (dTMP). This process of adding a phosphate group is called **phosphorylation** and is a necessary step in making dTTP (thymidine triphosphate), which is incorporated into DNA. Cells that contain the tk^+ gene can synthesize dTTP in this manner, but it is not the major pathway, so the cell does not *have* to use this mechanism. Cells have been discovered which are mutant in the tk gene, making them tk^- cells. These cells are unable to make thymidine kinase, which makes it impossible for them to make dTTP. They can be isolated as outlined in Figure 8-1.

Mutant cells that lack *tk⁻* can be isolated by feeding the cultured cells bromodeoxyuridine (BrUdr), a derivative of uridine that is also phosphorylated by *tk*. When BrUdr is phosphorylated by *tk*, it can be incorporated into the newly made DNA. This causes the new DNA to become susceptible to ultraviolet light. So when cells are grown on BrUdr and then subjected to ultraviolet light, the cells that have BrUdr in their DNA die. Those cells that have no *tk* (*tk⁻*) live just fine, although they are rare. Once these *tk⁻* cells are isolated, entire *tk⁻* cell lines can be grown (see Fig. 8-1).

Using HAT Medium to Screen for Transfected Cells

If normal cells are grown in a medium containing **h**ypoxanthine, **a**minopterin, and **t**hymidine (**HAT medium**), the normal synthesis of DNA is blocked by the aminopterin. Because thymidine kinase is present (*tk⁺* cells), the cells will grow in the presence of aminopterin because they can use the alternate pathway to make DNA. But if we try to grow *tk⁻* cells in HAT medium, they will not grow because they have no way to make DNA, since the normal pathway is blocked by aminopterin. The alternate (*tk*) pathway will not work, since the cells are mutant in this enzyme (*tk⁻*).

Let's take a *tk* gene from a normal cell and ligate (tie) it to a new gene that we wish to insert into a eukaryotic cell line. We then use either calcium phosphate or electroporation to allow the new DNA to enter *tk⁻* cells. Then, in the presence of the HAT medium, only the cells that have the insert will grow, because they now have the *tk* gene. This can be confirmed by analyzing the DNA and showing that the new *tk* gene is present in the cells that lived. This experiment is shown in Figure 8-2. So, now we can use the *tk* gene as a marker to tell us when we have transfected our cell line, since the cells that live should contain our new DNA insert.

By ligating (attaching) any new DNA to the *tk* gene, it should be possible to introduce any new DNA gene of our choice into mammalian cells. But it turns out that we don't even need to ligate the new DNA to the *tk* gene, since by merely adding the *tk* gene along with the gene to be inserted and a carrier DNA (e.g., hamster DNA) to the sample, the DNAs automatically blended with the *tk* gene (Fig. 8-3). It was found upon sequencing that,

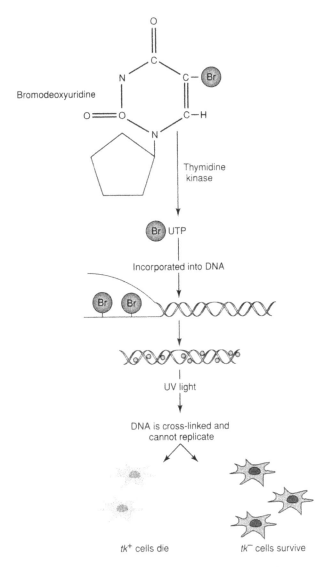

Bromodeoxyuridine

Thymidine kinase

Br UTP

Incorporated into DNA

Br Br

UV light

DNA is cross-linked and cannot replicate

tk⁺ cells die

tk⁻ cells survive

Figure 8.1 Screening using bromodeoxyuridine (BrUdr). Mutant cells that lack tk⁻ can be isolated by feeding the cultured cells BrUdr, a derivative of uridine that is also phosphorylated by *tk*. When BrUdr is phosphorylated, it can be incorporated into the newly made DNA. This causes the new DNA to become susceptible to ultraviolet light. So when cells are grown on BrUdr and then subjected to ultraviolet light, the cells that have BrUdr in their DNA die. Those that are *tk⁻* live just fine, although they are rare, because they couldn't put the BrUdr into their DNA. Once these *tk⁻* cells are isolated, entire *tk⁻* cell lines can be grown.

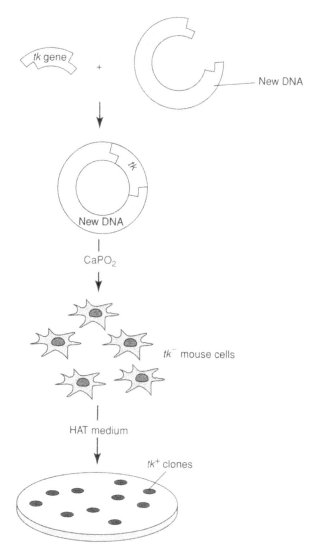

Figure 8.2 Screening cells using the *tk⁻* marker. First, the *tk* gene is attached to the piece of DNA to be inserted into a cell. Then the chimeric DNA is introduced into the *tk⁻* cell line by electroporation. Cells containing the *tk* gene will grow on the HAT medium, but all others will not grow.

although these various pieces were integrated randomly into the chromosome, the newly inserted DNA is physically linked by the cell to the *tk* gene that was added. So almost any new gene can be introduced into eukaryotic tissue culture cells simply by

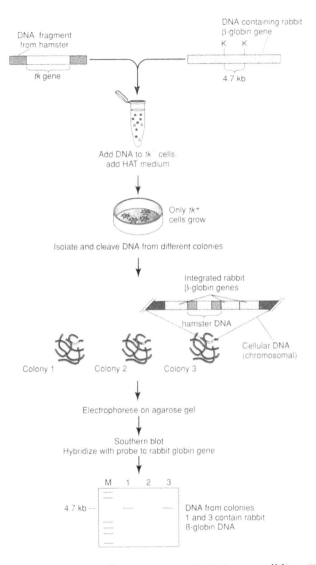

Figure 8.3 The process of inserting new DNA into a cell line. First, a DNA fragment containing the *tk* gene is mixed with a piece of DNA containing the DNA of choice (in this case a rabbit β-globin gene). By adding the DNA to *tk*⁻ cells (generally using electroporation) and growing the cells in HAT medium (**h**ypoxanthine, **a**minopterin, and **t**hymidine), only those cells into which the *tk* was inserted will grow. Culturing various colonies of the *tk*⁺ cells, the DNA from each is extracted, and a Southern blot is performed to identify those colonies into which the rabbit β-globin gene was successfully transferred.

including the DNA along with the marker and using calcium phosphate to incorporate the new DNA into the cell.

But using this thymidine kinase marker does not always produce the right change, because the new gene can be put almost randomly into the chromosome of the mammalian cells. It may or may not be in a region where it can be expressed into a protein. In addition, it may be placed in a position where it causes major changes in the mammalian cells themselves.

So, to get the new gene to be expressed, it is very desirable to add a promoter region to the gene and the marker. This combination of a new gene with a promoter region (and marker) attached is called a **construct**. This construct might look like that illustrated in Figure 8-4.

So, when the new DNA gene is put into a cell line, the gene will carry its own mechanisms to turn the new gene on and off. By using promoter regions that are stimulated into activity only by specific chemicals, the process becomes even more highly defined. Once the new DNA insert is in the cell line, we can add the specific chemicals that will stimulate production of message RNA form the specific gene we inserted. This will then be used to make the protein that was encoded in the new DNA gene.

The final product of all this is a mammalian cell culture that will be able to produce certain specific proteins, such as tissue plasminogen activator, which can be used to help fight blood clotting. As noted earlier, in many cases bacteria cannot make eukaryotic proteins, so the mammalian cell culture approach is essential.

VIRAL APPROACH

Mammalian cells can also be transformed using viral vectors (carriers). Initially, it was hard to find a viral vector that would work. However, a virus obtained from monkeys, SV40, was found to work well and has been the workhorse of animal cell transformation for many years. The SV40 genome is a small circular double-stranded DNA of about 5.2 kb (kilobases) with many restriction enzyme sites.

This SV40 virus will go through a lytic cycle in which it reproduces itself when infected into cell lines developed from the African green monkey. However, in mouse and hamster cell lines, there is no lytic infection, but the viral genome is inte-

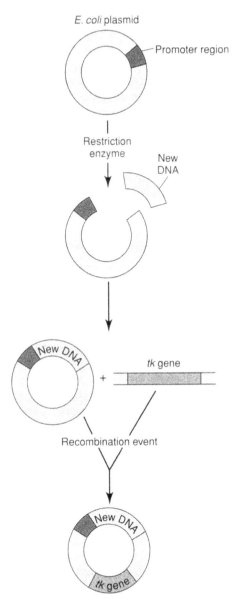

Figure 8.4   The making of a construct. To get inserted genes to turn on synthesis, a promoter region is placed next to the gene by cutting the plasmid next to the promoter region and inserting the new DNA gene of choice. The *tk* gene is added randomly into the plasmid as well. This entire unit can now be put into a mammalian cell, using electroporation, as shown in Figure 8-3.

grated into the host genome as in the lysogenic cycle of phage in bacteria. Often the SV40 sequences are rearranged in such transformed cell DNA. As with bacteriophage, there is a strict size limitation as to how much DNA the virus can carry. It also makes a difference as to where the insert is placed in the SV40 genome. Although a portion of the genome can be replaced, the size of the insert must be essentially the same size as the extracted DNA.

SV40 has been used successfully for a long time as a viral vector in many different applications. However, because of the size limitation and the problem that often occurs with rearrangements of the inserted DNA, other viral vectors have been sought.

Retroviral Approach

Recently, researchers have turned to **retroviruses** to deliver new DNA into mammalian cells (Fig. 8-5). Retroviruses have RNA (ribonucleic acid) as their genome and make DNA out of their RNA once they have infected a host cell.

The beauty of the retroviral system for our purposes is that the new DNA is placed in an exact place in the chromosome of the host cell and is generally readily activated. So if we were able to put our new gene into the retrovirus, it should be expressed as well. However, this is where it gets a bit sticky. First, retroviruses are often very infective to man and other animals. So it isn't a good idea to infect cells with an active viral agent that can infect man or animals. Second, there is still a strict size limit on the viral genome, and the addition of much new RNA is impossible.

Both of the latter problems have been bypassed neatly by taking major portions of the viral genome out of the virus and inserting the new RNA in that region.

As this viral genome is inserted into a cell, it is inserted into the host cell chromosome at the specific site where the virus would ordinarily insert its genome. Ordinarily, the virus would then go about making new viruses. However, if the construct we make replaces some or all the viral coat proteins, no new viruses could be made.

The major problem is to make these viruses that will infect cells properly, but not reproduce. The trick is to use **packaging cells**. These cells make the necessary viral coat and other

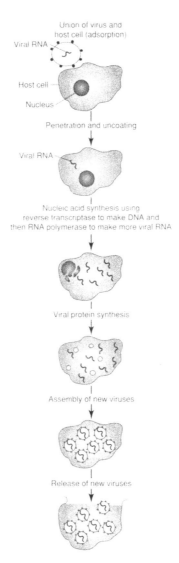

Figure 8.5 The life cycle of a retrovirus. As with the lytic pathway of a bacteriophage, a retrovirus attaches to a cell and the genome, in this case viral RNA, and is introduced into the host cell. An enzyme called reverse transcriptase first makes a DNA copy of the viral RNA. Then, RNA polymerase is used to make many more copies of the viral RNA. Some of the viral RNA is translated into proteins, and the remainder is used in the assembly of new retroviruses. In this case, the most cells may not be lysed, but will keep on producing and releasing retrovirus particles for some time.

proteins. So, by adding viral genomes to the packaging cell that has the genes for proteins deleted, the new genes will be inserted in that region and the packaging cells will then make new virus particles that contain a construct of the viral genome. This construct will contain the new gene and promoter region. The virus that is produced is a **crippled virus** in that it cannot reproduce itself, but is nonetheless infective to other cells (Fig. 8-6).

Packaging cells are designed to produce tremendous numbers of virus particles in about 2 days. The crippled viruses can then be used to transfer their particular gene inserts to various mammal cell cultures. When a virus carries new DNA into a cell, it is called a viral **vector**. A vector is any device or agent that can be used to insert new DNA into a cell. So, not only viruses, but phage and plasmids are vectors.

The retroviral approach is a powerful approach to use to insert new genetic material, since the crippled retroviruses place the new DNA in active regions of the host cell chromosome, which allows expression of those new genes. The retroviral vector approach has been refined over the years to produce better and more specific packaging cell lines, more specific viral vectors, and increased production of the virus particles in the packaging cells.

The retrovirus of choice in recent years has been the mouse leukemia virus (murine leukemia virus, MLV), which has been crippled and modified as just described. Unfortunately, like most retroviruses, this virus can only infect cells if they are dividing. This works fine in mammalian cell cultures, but does not work well for gene transfer to living tissues in animal cells that may not be actively dividing. It is interesting that very recent success has been reported using the human immunodeficiency virus (HIV) as a crippled virus vector for nondividing cells. This approach will be discussed more fully in Chapter 11.

We can see that the possibilities of transfecting animal cells are numerous and fall essentially in the same patterns as the transformation of bacterial cells. As noted earlier, if mass production of a certain protein, such as tissue plasminogen activator, is needed, because bacterial cells cannot be used for this purpose, transfected mammalian cells are extremely useful and important to provide this important protein. The method of choice to transform mammalian cell cultures is the use of a viral vector.

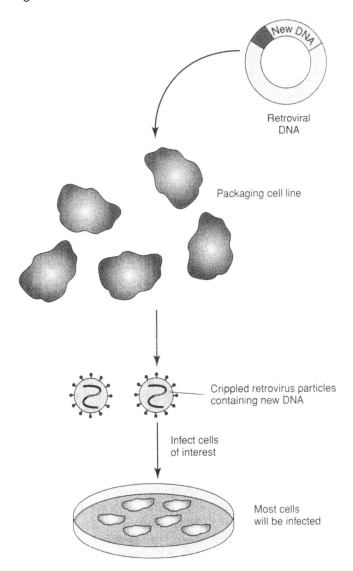

Retroviral
DNA

Packaging cell line

Crippled retrovirus particles
containing new DNA

Infect cells
of interest

Most cells
will be infected

Figure 8.6 Making crippled retrovirus particles that can infect other cell lines. Initially, deletions are made in some of the protein genes of the retroviral RNA. In the place where the deletions occurred, new DNA containing a promoter can be inserted. The DNA construct is then inserted (by electroporation) into packaging cells. Packaging cells rapidly encapsulate the viral RNA genome into virus particles and release them. The viral RNA genome still lacks some viral protein genes and contains the new DNA. These crippled virus particles are then capable of infecting a new cell line, but cannot reproduce themselves.

cDNA LIBRARY

We would like to place a human gene for a specific protein into a mammalian cell culture. To do this, it is highly desirable to make a library of human cell DNA so that it will be available to sort through and look for certain genes. In Chapter 6, we learned how to make gene libraries for a pancreas cell (which is a cell derived from mammals) by cutting it up into small pieces and putting these pieces into bacterial cells. As noted, gene libraries are wonderful sources if we want to get a lot of a certain gene (such as the insulin gene found in pancreas cells)

But in Chapter 3, we learned that the chromosomes from higher-order (eukaryotic) cells had many pieces in them that were not used as genes (introns). These introns had to be removed from the pre-messenger RNA before it could be used as message. So, it would be highly desirable to have a library containing the messenger genes only, without all of the introns present, as were present in the library we previously discussed.

Actually, this can be done quite well. All eukaryotic cells use mRNA (messenger RNA) molecules that have a poly-A tail. If we take a eukaryotic cell line and disrupt the cells, it is possible to "fish out" the mRNA molecules by using an oligo-T column. An oligo-T column is filled with tiny glass beads to which are attached pieces of DNA containing only T (thymidine) residues. These stretches of T–T–T–T–T will find the A–A–A–A–A tail of the mRNA and stick to it with hydrogen bonds in a complementary way, as shown in Figure 8-7.

The poly–A–containing mRNA molecules can be washed off the oligo-T column using a salt solution. These mRNA molecules contain the genetic codes of all of the proteins that the eukaryotic cells were making when they were harvested. The advantage of using these mRNA molecules rather than the DNA genes is that all unnecessary information contained in the DNA in the form of introns has been removed, leaving only the genetic messages.

Using reverse transcriptase (an enzyme that makes DNA from RNA template), a DNA copy of the mRNA can be made. By using a poly-T primer (a short piece of DNA complementary to the poly-A tail of the mRNA), the DNA complementary to the mRNA can be made, which is called complementary DNA or **cDNA**. By adding poly-T and reverse transcriptase to the mRNA molecules isolated from a given cell line, a multitude of cDNA molecules will be made, corresponding to the various mRNA molecules in the cell line (Fig. 8-8). These cDNA

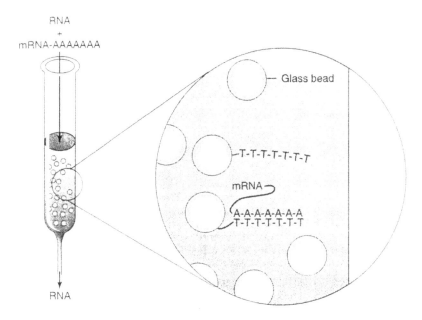

mRNA-AAAAAAA is bound to oligo-T column

Figure 8.7 Purification of messenger RNA (mRNA). mRNA from eukaryotic (higher-order) cells always has a long string of A's on the 3′ end (a poly-A tail). Since this would be complementary to a series of T's, a column is made containing beads with a long string of T's (oligo-T) attached. When a solution containing mRNA is poured through the column material, everything goes through but the mRNA, which sticks because of the complementary base pairing between the A's at the 3′ end of mRNA and the T's attached to the column matrix. The mRNA can then be washed off with a salt (e.g., NaCl) solution.

molecules contain the genetic message of the various proteins that are being made by that particular cell line.

The cDNA strands now can be inserted into plasmids by adding the appropriate end groups and tying them into the plasmids exactly as outlined in Chapter 5. The modified plasmids are then inserted into bacterial cells and the culture of cells grown, giving a **cDNA library**.

Screening cDNA Libraries

The screening of cDNA libraries can be accomplished in exactly the same way as we did with the gene libraries, as discussed in

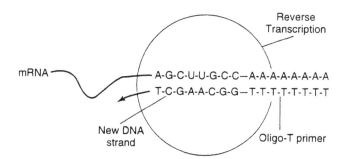

Figure 8.8   Diagram showing how reverse transcriptase works. First, we need an RNA template and a primer. The primer is complementary to the 3′ end of the template. If, as shown in this example, the template is messenger RNA (mRNA), the 3′ end is a series of A's, so the primer can be an oligo-T (a series of T's). Upon adding the primer, reverse transcriptase, and the necessary ATP, GTP, CTP, and GTP, reverse transcriptase synthesizes a strand of DNA complementary to the RNA template.

Chapter 6. In this case, the DNA probe used would be complementary to a portion of the new gene sequence. This would identify any cDNA that contained that sequence. As previously noted, the advantage of this library over a gene library is that the cDNA library contains genes that can be transcribed directly into mRNA without further processing. This makes them ideal for inserts into cells. By isolating a specific gene from a cDNA library, adding a promoter region to it, and inserting it into mammalian cells and culturing them, immense amounts of that particular protein will be made by the cells. The cells can then be harvested and the protein purified.

APPLICATIONS

We now have at hand a neat arsenal of tools to use to harness cells to do work for us. We can isolate particular genes from gene libraries and then, using the techniques we have discussed in the previous few chapters, insert these new pieces of DNA into the appropriate cells of our choice. It is important that these new pieces of DNA contain a gene for a protein we wish to make.

This approach has been used for many different proteins and is being pursued for many more. Table 8-1 lists a few of the

TABLE 8.1   Some Proteins Made Using Genetic Engineering

Protein	Use
Human insulin	Treatment of diabetes
Human growth hormone	Treatment of growth deficiencies
Interferon	Treatment of viral diseases
Tissue plasminogen activator	Dissolves blood clots
Erythropoietin	Stimulates red blood cell production

proteins that have been made using genetic engineering techniques, which are now commercially available. Insulin, which is used to help control diabetes, has been made primarily from bacterial sources. Tissue plasminogen activator, a protein used to dissolve blood clots, has been made using both bacterial and mammalian cultures.

Human interferons, at least some of them, have been made using both bacterial and tissue cultures. Although these are known to inhibit viral disease, their exact mechanism is not well understood at this time. Some of the interferons are glycoproteins (have sugars attached), so they must be made in mammalian systems.

All of these examples are proteins that are overproduced by genetically engineered bacterial or cell culture approaches, which allow us to produce large amounts of proteins that can be administered to prevent or control certain diseases. There are two major advantages in using this approach, rather than trying to isolate these substances from naturally occurring sources.

First, the amounts that can be made are almost unlimited; yet, the cost is but a fraction of what it would cost to isolate the protein from a natural source. Second, in all cases we can use the human protein.

For instance, insulin was isolated from pigs before it was genetically engineered into a bacterial strain. As a result, it was quite expensive. But there were also a number of people who had adverse reactions to porcine insulin. So, with the development of genetic engineering techniques, both problems were addressed.

In the case of human interferon, it took over 5000 gallons of blood serum to produce 1 g of interferon before it was genetically engineered in the early 1980s. This meant that the cost of treatment to patients was almost prohibitive. Now, it can readily be produced in gram amounts for a fraction of the cost.

These and other examples illustrate the power of the genetic engineering approach to supplying proteins that are critical in medical treatments.

CONCLUSIONS

We need to emphasize that thus far, we are really establishing "protein factories" using transformed cell lines. Although it is less costly if bacteria can be harnessed to provide various proteins, some proteins have sugar units attached or other complexities and need to have additional components, which bacteria cannot readily provide. In these cases, animal cell lines are used. In Chapter 11, we will talk about many of these applications. This approach is substantially more expensive than bacterial cultures, but it is still markedly less expensive than extracting proteins from other natural sources.

So far in this book, we have developed an arsenal of techniques that can be used to put new pieces of DNA into either bacterial or animal cells. In each case, we have learned ways to put the inserts into desired locations in plasmids or genomes. We have also developed ways to find them once they are inserted. This is done by using markers to screen for the cell lines that have been transformed. The final proof is whether the transformed cell lines actually make the new proteins whose genes have been inserted.

SUMMARY

Although DNA can be inserted into mammalian cells using either calcium phosphate or by electroporation, the placement of the DNA at the right spot in the chromosome of the cell is important. We monitor whether our new DNA insert is present in the cell using a tyrosine kinase (*tk*) marker. This marker is a little like the antibiotic-resistant markers used in bacterial cells, in that when it is present, cells grow on a special culture medium. By using cells that don't have the *tk* gene and then adding the *tk* gene with our new DNA, we can screen for the cells that have the tyrosine kinase marker and the new DNA. The problem remains that even when the new DNA is inserted, it may not go to an active region of the mammalian cell's genome and would not be useful.

Viral vectors were developed to insert the new DNA into a correct region of the mammalian cell genome. By placing a promoter region with the new gene, the gene can be turned on with specific chemicals added to the culture medium. In this way, mammalian cells can be cultured, the genes turned on, and specific proteins manufactured for medicinal use by man.

To obtain the desired genes, we need to make a library of the specific genes used to code for the proteins. By obtaining messenger RNA (mRNA) molecules from growing human cells and reverse transcribing the mRNA into complementary DNA (cDNA), a library of these cDNA pieces can be made by placing them into bacteria. We can then screen the bacterial cDNA library for a desired gene, isolate that bacterial strain, extract the cDNA fragment, and insert it, along with a promoter, into the mammalian cell line to produce the protein we desire to be made. This approach now is routinely used for a number of proteins, which we will discuss more fully in Chapter 11.

9

GENETIC ENGINEERING OF PLANTS

WHAT YOU WILL LEARN IN THIS CHAPTER

- Some historical methods of plant breeding and engineering
- How to obtain and culture plant cells having no cell walls (protoplasts)
- How to put new DNA into plant cells
- How the leaf-disk technique is used to transfer new DNA into plant cells
- How to use viruses to transfer new DNA into plant cells
- How antisense nucleic acids are used to engineer plant functions

So far we have concentrated almost entirely on the ways and means to genetically alter bacteria and animal cells, but we have said nothing about whole organisms, such as plants or animals. Plants are becoming a very fertile area in which to perform genetic engineering. For instance, a new variety of tomato was introduced recently—the Flavr-Savr™. This tomato can be picked almost ripe from the vine, then shipped without refrigeration and still remain firm and unspoiled on the grocer's shelf for over twice as long as the typical green-picked tomato. The Flavr-Savr™ was genetically engineered to have these qualities.

How can we genetically engineer plants? Plants are composed of cells, so we might assume that the same methods we have used with bacteria will work here. This is true in principle, but plant cells all have a heavy cell wall that makes them pretty impenetrable to most of the techniques we have learned thus far. In addition, plant cells generally do not contain plasmids or other nice features we have worked with up to now.

HISTORICAL METHODS

There are several genetic engineering approaches to take with plants. And they have lots of advantages. Plants are easy to cross and often can be bred asexually. For thousands of years, plants have been grafted and cross-bred, but in the last two decades or so, many sophisticated variations have been introduced.

In the past, when the goal was to introduce traits such as disease resistance, two plant lines were sexually crossed to give first-generation hybrids. These hybrids were then crossed with the parents until the desired traits were present. This process was often tedious and was restricted to initial crossing of species that were sexually compatible.

Technologists who work with plants have one tremendous advantage over those who work with other organisms. It turns out that many plant cells are **totipotent**, even mature plants. Such cells can be grown in a medium and then induced to produce plants from single cells. A totipotent cell is one that contains the full complement of DNA, all of which is able to be used.

In recent years, scientists have found ways to cause the cells of plants to grow in culture media, much like bacteria grow. Certain cells can be isolated from plants. When they are placed in the right growth medium containing certain growth hormones, they continue to divide and grow individually. These cells then can be treated with other hormones, which cause them to differentiate into whole plants or portions of plants. This redifferentiating process can only happen when the cells have not been cultured long. That is, great, great, great granddaughter cultured cells are less likely to redifferentiate than their grandparents. It is not certain what is lost as the cells grow older in culture, but it is known that early cultures work better and late cultures hardly work at all. This kind of tissue culture is very common in research and commercial laboratories today (Fig. 9-1).

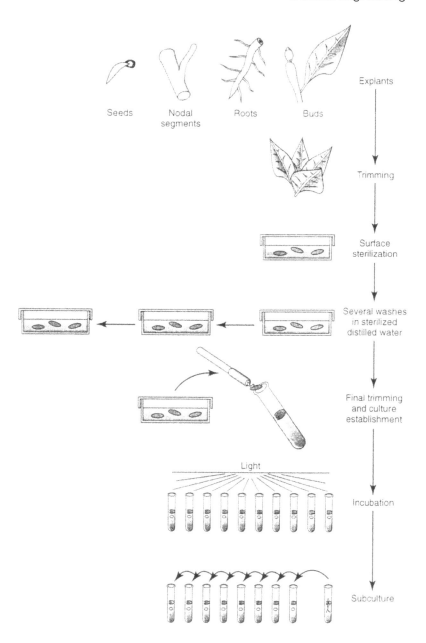

Figure 9-1 Overview of the way in which tissue culture can be set up. Starting with portions of plants, sterilizing, washing, and trimming take place, after which the pieces of plant are placed in growth medium. The cells then multiply and sometimes even develop into plantlets. They are subcultured (transferred again) to produce tissue in culture. Separation of cells into a cell culture can also take place.

From such tissue culture processes, we can obtain many clones of the original cells, each of which can then be redifferentiated and placed in culture. Again, when appropriate hormones are added, whole plants grow. Each plant is derived from a single cell and are clones of one another. In this way, florists have grown rare and exotic flowers in great numbers. For instance, most commercial orchids are grown from cloned plants.

As noted earlier, it *should* be simple to modify the plant chromosome with a novel gene, which will then allow a new transformed cell line to come about. But it is not simple at all, owing to the thick cell walls, which make insertion of new genetic material very difficult (Fig. 9-2).

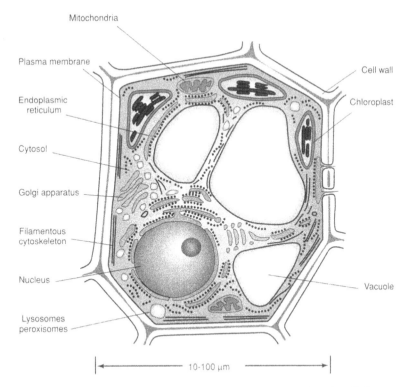

Figure 9-2 Diagram of a thin section of a generalized cell from a higher plant. Note especially the thick cell wall, which is not found in animal cells.

PROTOPLAST CULTURE TECHNIQUES

To try to get around the cell wall problem, scientists have learned how to remove the cell walls from plant cells, leaving a plant cell without a cell wall. This cell without a cell wall is called a **protoplast**. The protoplast is much more vulnerable to insertion of new DNA.

The source of cells from which to make protoplasts can be a cell culture, callus, or a sterilized leaf (Fig. 9-3). **Callus** is just a

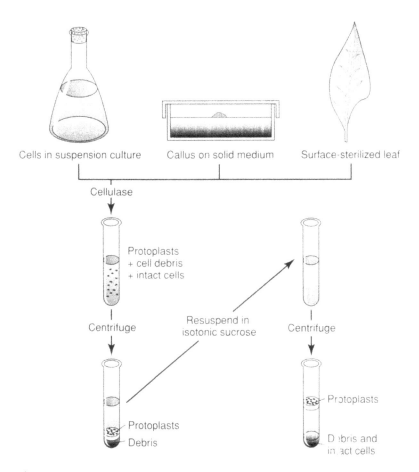

Figure 9-3 Preparation of protoplasts. Protoplasts are plant cells without the thick cell wall. Cells can be obtained from any source. The addition of cellulase destroys cell walls, after which two centrifugations separate the protoplasts. The second centrifugation step is done in sucrose—a denser medium than water—so the protoplasts float.

random mass of cells that adhere to one another in a culture medium. Using cells from any of these sources, on the addition of **cellulase**, an enzyme that destroys cell walls, the remaining cells without walls are the protoplasts. These are then purified and isolated by centrifugation.

If these protoplasts are placed in the proper growth medium, they will propagate themselves wildly, and in so doing clone themselves. While they are in the protoplast stage, they can be kept separate or induced to fuse with each other or with protoplasts from other plant cells. Using this approach, we can make hybrids of new and unusual sorts from otherwise sexually incompatible cells. For instance, a "pomato" is made by fusing tomato and potato protoplasts.

By using calcium ions or ethylene glycol, protoplasts can be induced to take up other components in the growth medium, including pieces of DNA ordinarily unable to penetrate the plant cell wall. Although protoplasts can get new DNA directly in this way, the product of that transformation is not always predictable, because the newly inserted DNA may not find its way to a productive point in the protoplast chromosome. Although the protoplast culture techniques show promise, they have not yet been useful in producing viable new hybrid species that can reproduce without considerable crossing of the newly formed plant with other plants.

GENETIC ENGINEERING APPROACHES

What is really needed is to have the ability to introduce new genes into plants in a site-specific manner. First, as before, it is important to obtain the useful genes in sufficient amounts to transplant into another organism. Second, it is necessary to insert these genes where they can be expressed when needed.

The first step in genetic engineering is now easily done. We can use the techniques we have previously learned and grow immense amounts of DNA—even plant DNA—in bacteria to give plant DNA libraries. These DNA or complementary DNA (cDNA) libraries can then be screened to identify and isolate those bacteria that have the needed DNA in their plasmids. These bacteria can then be grown and the plasmids can be harvested from the bacteria using the methods we have discussed.

Agrobacterium tumefaciens Technique

Putting new plant DNA into the plant cell in a position so that it will be expressed properly is another story. A method of doing this has recently been found and developed and is now commonly used.

A certain strain of bacteria from the soil, *Agrobacterium tumefaciens*, can infect certain types of plants. When this happens, *crown gall tumors* form (Fig. 9-4). These tumors consist of plant cells that have gone wild—much like cancer cells in humans. It was found that agrobacteria insert a plasmid into the plant cells. This plasmid is called the **Ti plasmid**, which stands for "tumor-inducing." When the agrobacteria infect the plant, the plasmid enters the DNA of the plant cells. This causes the plant cells to form crown gall tumors. These tumors can be excised, put into culture, and grown as callus there.

This scenario is similar to what we found in bacteria, in which a plasmid was inserted into a bacterium and transformed the bacterium in a specific fashion. We used the plasmid to transfer DNA of our choice to the bacteria, using the plasmid as a carrier. In a similar fashion, it should be possible to use the Ti plasmid as a carrier for DNA of our choice, which can be used to transform plant cells.

Because of its size, the Ti plasmid is difficult to manipulate, but it has been possible, using several different methods, to insert new genetic material into the tumor-inducing region of the plasmid. Initially, the genes for tumor induction are removed from the Ti plasmids, leaving this region available for new DNA. Insertion of this modified plasmid no longer causes the crown gall tumors to form, but does allow the new DNA to be inserted into the plant cell chromosome. This approach has been successfully used and is now in wide use to insert new genes into plants.

Use of Leaf Disks With Agrobacteria

A technique of choice to place new DNA in plant cells is to use leaf disks. Small circles (about 1 cm in diameter) of leaves are cut and infected with an *Agrobacterium* culture that contains a genetically modified plasmid—one that has the new DNA in it. To find out whether the new DNA has been inserted, an antibiotic resistance gene (e.g., a kanamycin-resistant gene) is joined to the new gene as a marker and is inserted into the plasmid (Fig. 9-5).

The plasmids are then inserted into the *Agrobacterium*, following which the leaf disks are infected with this genetically modified organism. The leaf disks are then grown on medium

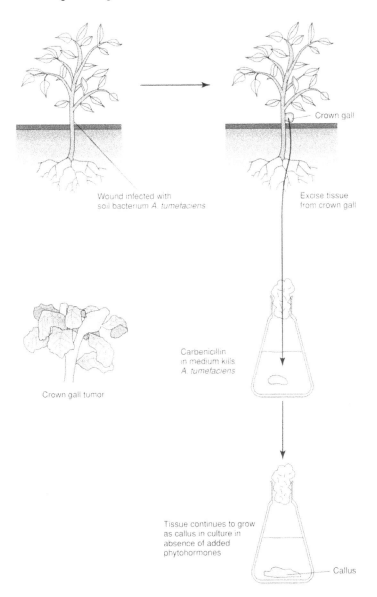

Crown gall

Wound infected with
soil bacterium *A. tumefaciens*

Excise tissue
from crown gall

Crown gall tumor

Carbenicillin
in medium kills
A. tumefaciens

Tissue continues to grow
as callus in culture in
absence of added
phytohormones

Callus

Figure 9-4 Infection of plants by the soil bacterium *Agrobacterium tumefaciens* causes a tumor to grow on plants. this tumor, called crown gall, can be removed and the tissue extracted and put in culture in a solution containing an antibiotic to kill the *A. tumefaciens*. In culture medium, the crown gall tissue grows into a callus that will produce cells that can be grown in tissue culture as well.

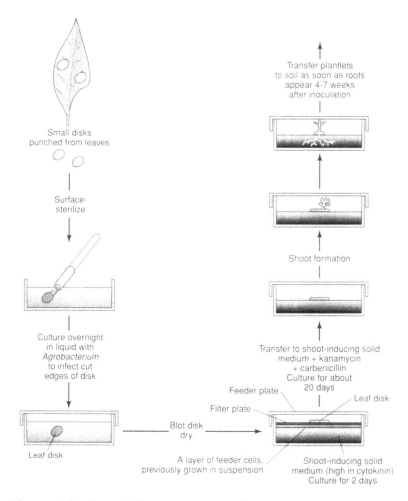

Figure 9-5 Genetically engineering a plant. Small disks are taken from the leaves of a plant we wish to genetically alter. The agrobacteria have previously been genetically altered to contain the gene we desire. This gene is placed in the Ti plasmids just as we did with bacteria. Once infected with the *Agrobacterium*, the leaf disk is then put through a series of growth steps to produce the new plant. Note that the process is much more lengthy than the process in bacteria because of the growth times.

that contains an antibiotic, such as kanamycin. The cells that are transformed by the *Agrobacterium* plasmids will be kanamycin-resistant and will grow spontaneously in the medium. The nontransformed cells will not grow.

Note that the leaf disk approach is only useful for some plants and is not universally applicable to all plants, especially cereal plants, such as wheat, soybeans, and other grains.

Inserting New DNA with Viruses

It is also possible to use viruses to place new DNA into plant cells, as with mammalian cells, which we discussed in Chapter 8. Two DNA-containing viruses, cauliflower mosaic virus and gemini virus, have been used for this purpose. However, every virus has limitations with respect to the kind of plants they will infect, the amount of inserted DNA they can carry, and exactly where they will insert that DNA in the plant cellular chromosome. Efforts are ongoing with these vectors, and some successes are being realized. The main advantage is that these viruses can insert new DNA into cereal plants (such as wheat, oats, and rye), which agrobacteria cannot do.

Because most plant viruses are RNA (ribonucleic acid) viruses, it makes the process of viral transfer of genetic information somewhat more difficult than with DNA viruses. Brome mosaic virus and tobacco mosaic virus are prime RNA virus candidates at this time. The major problem is engineering the insert to be placed in the viral RNA. RNA is more fragile than DNA, which can be readily synthesized or obtained from other sources. Although RNA can be synthesized, the process is much more expensive and much less accurate than DNA synthesis. Generally, RNA is made using RNA polymerase, an enzyme which will make the RNA from a DNA template. This added step produces an additional hurdle for the RNA virus approach.

Tomato golden mosaic virus has recently been developed as a vector. It has two single-stranded DNA molecules in its genome. These naked DNAs are invective to plants. By inserting new DNA into one of these and inoculating into plants with this virus, viral infection spreads the new DNA into the plant cells.

The purpose of using the viral approach is to get the new DNA inserted into active sites within the plant chromosomes. Unfortunately, at this time no single virus works on all plant species. Thus, there is still an ongoing search for the best viral vectors to use in the engineering of plant cells.

Direct Methods

Other direct approaches have been made to insert DNA directly into plant cells through the tough cell walls. Although these methods lose the specificity of the plasmid or viral approaches (the ability to place the DNA at a strategic site in the plant cell chromosome), they are very convenient and have worked in numerous instances

Insertion of DNA directly with a projectile gun has seen moderate success. High-velocity microprojectiles deliver nucleic acids into intact cells and tissues. Although many names are used, this process is generally known as the **biolistic process**. In this approach, a heavy metal—often gold or tungsten—is coated with a DNA plasmid or a mRNA (messenger RNA) and then literally shot into the cells to be transformed. Figure 9-6 compares this projectile approach with the *Agrobacterium* approach to insert new DNA into plant cells.

Most often used with plant cells, this approach is also useful with other cells, including mammalian cells and even whole animals. This approach is direct and does work, but is still being improved and will likely be most useful for specialized application. This is literally a "shotgun" approach, since the new DNA is not directed to a specific site in the plant cell chromosome.

Another direct approach is to microinject plant cells using a syringe. This can be done easily, although it is technically demanding. However, up to 100 cells per hour can be microinjected by a competent technician, with a high percentage of success. Again the time needed and cost involved make this technique relatively unlikely to be a commercial success in most cases.

A Different Approach—The Antisense Approach

To this point we have been discussing ways in which new DNA can be inserted into plant chromosomes. In this way, genes for better or different proteins can be added to the plant cells. But what if we wanted to turn off a gene for a certain enzyme?

A way to do this is to genetically alter the cell so that it produces a piece of RNA that effectively blocks the transcription or translation of an undesired protein. It is called "antisense," because it is complementary (antisense) to a "sense" strand of nucleic acid found in the DNA or mRNA.

For instance, in 1988, a researcher reported finding a way to switch off the gene that makes harvested tomatoes mushy, allowing vine-ripened tomatoes to be delivered to your door. Polygalacturonase (PG) is the enzyme involved in the softening of ripening tomatoes. Biologists blocked synthesis of this enzyme by inserting an "antisense" gene into the plants

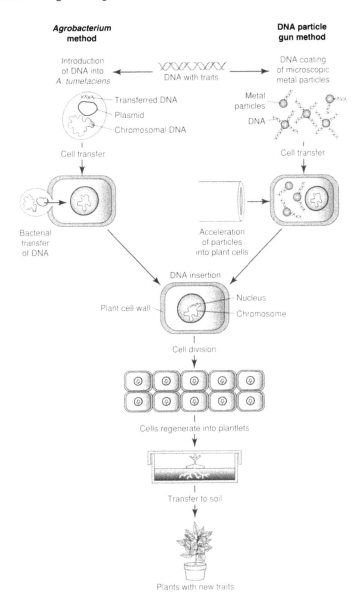

Figure 9-6 Two common methods for genetically alter plants. The *Agrobacterium* method is described in the text and in Figure 9-5. The *Agrobacterium* particle gun method uses metal particles that have been coated with the DNA to be inserted. These particles are literally shot into plant cells, penetrating the cell walls. The DNA is taken up by the chromosome of the plant and, in some cases, genetically transforms the plants, using the new genes that have been inserted.

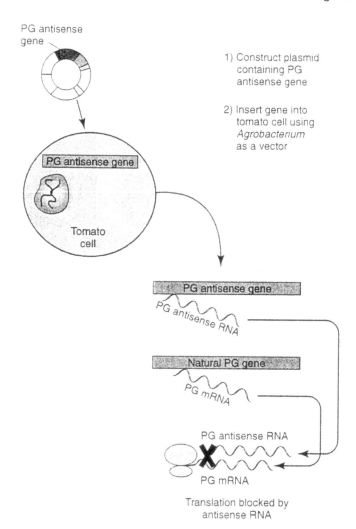

Figure 9-7 The antisense approach. Although it carries a curious name, this powerful approach has already been used for a commercial product—vine-ripened tomatoes that don't go soft. A polygalacturonase (PG) gene is inserted into a tomato cell line. The PG gene makes a piece of RNA that is complementary to the messenger RNA (mRNA), which makes the PG enzyme. When the complementary (antisense) RNA binds to the mRNA, the ribosome cannot translate the message, so the PG enzyme is not produced. It is the PG enzyme that makes tomatoes go soft.

(Fig. 9-7]. This gene blocked the site at which transcription of the PG gene would normally begin. The result is that vine-ripened tomatoes reach the market firm and fresh!

APPLICATIONS

Engineered bacteria are now used as antifrost devices for strawberries to allow year-around strawberry production. Although this is not a process wherein the bacteria transfer genes to the strawberry plants, the bacteria themselves are engineered to produce a product that helps protect the strawberry plants. This has not been as successful as hoped, but it does allow additional growth periods for the strawberries.

Many plants now being tested contain added genes that confer disease resistance to the organism. Cotton plants resistant to cotton bollworms, corn plants resistant to corn borers, tobacco, tomato, and grape plants resistant to viruses that infect them all have been tested successfully and should soon be on their way to the market. Inserts have been made using *Agrobacterium* organisms as well as other vectors and have been incorporated as part of these strains. The major problem in this development is the release of genetically engineered organisms into the environment. But, there are inherent problems.

One approach has shown potential problems with gene manipulation: the engineering of glyphosphate resistance into various crop plants. Glyphosphate is the active ingredient in some herbicides (such as Roundup ® and Tumbleweed ®). By incorporating glyphosphate resistance into crop plants, the herbicides can be used when the crops are growing, without damage to the crops. But Mother Nature has found a way to transfer the genetic inserts from the plants into some weeds, which are now resistant to the herbicides. Prospective hazards of genetic engineering approaches are more fully discussed in Chapter 13.

CONCLUSIONS

All the genetic engineering techniques discussed here show promise, and much effort is directed toward modifying plants using all of them. The *Agrobacterium* approach is by far the most common at present. The growing activity in plant biotechnology is most likely due to the enormous market potential, the lack of severe governmental regulations such as those governing the manipulation of human genes, and the natural ability of plants to do a lot of genetic engineering on their own. We can look forward to tremendous advances in the future in this area of genetic engineering.

SUMMARY

Plant cells present more problems to the genetic engineer than do animal cells, mostly because they are encapsulated in a heavy cell wall, making transferring new DNA a challenging process. One approach is to remove the cell wall entirely. The resulting cells without walls—protoplasts—can receive new DNA directly through the cell membrane.

By using bacteria, plants can be infected when the bacteria insert plasmids into the plant cells. By placing selected DNA into the plasmids of these bacteria, this new DNA can be inserted into the infected plant cells, becomes part of the plant cell chromosome, and can provide resistance to certain diseases or be used to produce new proteins.

Viruses can also infect plant cells. By putting additional genetic material into plant viruses, these viruses can be used as the means to transport the new genetic material into plant cells. The advantage is that viruses are careful to put the new DNA in active sites in the plant cell chromosome.

Other methods have been used to direct plant cell activity. A successful approach is to put in a short piece of new DNA, and, when it is transcribed into RNA, the new RNA blocks the translation of some protein.

10

EMBRYO TRANSFERS AND CLONING OF ANIMALS

WHAT YOU WILL LEARN IN THIS CHAPTER

- How fertilization normally occurs
- How eggs can be fertilized in artificial (*in vitro*) circumstances
- How new nuclei can be transplanted into other cells
- How new genes can be inserted into egg cells to make transgenic animals
- What a "knock-out" mouse is
- How "pharm" animals can be developed and used
- How genetic engineering can be applied in other ways in the agricultural world

With our arsenal of tools, we are now prepared to engineer whole animals. How do we go about doing this? We must first learn a bit about the way animals start their existence.

THE BEGINNING OF ANIMAL LIFE

Initially, an egg is fertilized by a sperm. Each of these germ cells carries half of the chromosome complement of the mature cells.

When fertilization occurs, the egg for a short time has two **pronuclei** (each containing half of the total chromosomes)—one from the sperm and one from the egg. These two pronuclei combine to form the cell's nucleus and give the cell the full complement of DNA. A fertilized egg is called a **zygote** and is said to be **totipotent** because it is not specialized and can give rise to an entire functioning organism. All portions of the DNA are available for use by the cell. As this cell continues to divide, a time comes when the daughter cells are no longer totipotent, and **differentiation** begins. Differentiation is the process by which genes are selectively expressed to produce specialized cells.

Zygotes quickly divide repeatedly, forming groups of cells of 2, 4, 8, 16, 32, and more. During this early stage, the cells are all totipotent. Ultimately, they form a **blastocyst**, which is really a large group of the rapidly dividing cells contained in a membrane. The individual cells in the blastocyst are called **blastomeres**.

In the early stages of division, it is possible to separate the individual cells to obtain multiple copies of the same cell. However, once they are separated, these cells are very difficult to sheath together in a blastocyst. Nonetheless, it can be done. For instance, scientists in Scotland reported that they took cells from an embryo, grew thousands of individual copies in the laboratory, and then used these copies to produce a number of cloned sheep from ewes.

In vitro *Fertilization*

In a related process called *in vitro* (outside the living body) fertilization, a blastocyst is transferred to a surrogate mother. This process of fertilization is becoming more common in humans and animals. *In vitro* fertilization does not necessarily involve manipulation of the gene, but such transplants can occur after the gene has been manipulated. This approach has become an important technique in animal breeding and has substantial potential with humans as well. Techniques have recently been developed to separate male sperm from female sperm, allowing the animal clones to be sexed in advance. Such a process should provide tremendous advantages to both dairy and beef farmers.

CLONING ANIMALS

When we used the term "clone" before, we were referring to identical bacteria that all had the same DNA. The same general definition can be applied to animals. When two animals have identical chromosomal material, they are clones. Identical twins fall into this category, but we generally don't use the term for human twins.

When we want to make clones of animals, we go through the following process. First, we separate the cells of the animal to be cloned in the early division stage. Then these individual cells can be grown to the blastocyst stage and inserted into surrogate mothers. The resulting progeny would be clones because the DNA within the cells making up the progeny is identical.

Putting New Nuclei in Other Cells

Suppose we wanted to have a white mouse give birth to black progeny. This can be done by taking the fertilized egg from the white mouse and removing the two pronuclei. Then, if we had early blastomeres from a black mouse zygote, we could remove the nucleus from one of the blastomeres and insert it into the fertilized white mouse egg cell that lacked the pronuclei. The fertilized egg can then be grown to the blastocyst stage and inserted into the mother again. This could be done many times with many fertilized egg cells, and the resultant black offspring all would be clones of each other. Even though the fertilized egg cells came from white mice, the offspring all would be black. These are **transgenic** mice, because the white mothers have given birth to offspring with new DNA in them.

The first experiments using fertilized egg cells with new nuclei were done in 1952 with frogs' eggs. These eggs are very large compared with most other eggs and are therefore fairly easy to manipulate. It was found that if nuclei were removed from zygotes that had not gone through too many divisions, cloning was direct and easily accomplished. However, as the cells became progressively more differentiated, the nuclei lost their capacity to replace the nucleus of the fertilized egg. In fact, adult frogs were not able to donate viable nuclei at all.

Scientists have attempted similar experiments with some degree of success in larger animals such as mice, rats, sheep, goats, and others. If the nuclei are removed from a very early-stage blastomere of a donor animal and then inserted into a fertilized, denucleated egg cell from a receptor animal, the egg cell containing the new DNA could be made to live by growing it in culture to the blastocyst stage and then inserting this blastocyst into the uterus of a receptor animal. This is difficult, time-consuming, and expensive, so this effort has been used mainly for research purposes. If the efficiency and reproducibilty can be improved, cloning may become useful in animal breeding.

Note that it is now possible to grow a goat from zygote to birth completely outside the uterus. Although this approach is too expensive to be commonplace, it could take place routinely in animal breeding programs in the future. However, the science fiction specter of rows of developing identical children on shelves in an incubation facility, though technically possible, has enormous ethical problems in the case of humans.

Tradition has held that adult animal cells cannot be used as donors of genetic material because their cells have different-iated. However, in 1997, a sheep—one embryo from 277—was cloned from cultured udder cells from a 6-year-old ewe. This experiment shocked the genetic world, since previous efforts had always failed. To do this, researchers starved the udder cells, forcing most of their genes to enter into an inactive phase. Then, when the udder cell nuclei were transferred to the eggs, the "inactivated" nuclei were activated and one of the 277 produced a healthy, living animal—Dolly.

Needless to say, the results with Dolly have caused a storm of experimentation as well as controversy, especially regarding the idea of human cloning. The ethical considerations of human cloning are enormous. Scientific societies and numerous other political and religious groups have spoken out on this issue. Scientists have agreed, in general, that further cloning experi-ments will not be done on humans. Still, reports surface here and there of individuals or private concerns offering a cloning service for humans.

It is too early to predict all that will emanate from Dolly's emergence on the scene, but the future of cloning appears secure as protocol to develop identical cell lines for specific purposes.

GENETIC ENGINEERING OF ANIMALS

Although such mass cloning of animals is now a reality, the process of inserting new genetic material into these animals is another matter. So far, we have only dealt with transferring nuclei from one cell to another. Is it possible to use some of the techniques we have learned with bacteria, animal, and plant cells and apply these techniques to changing the DNA in an entire animal? The quick answer is "yes," but it is not straightforward.

In principle, the approach is to get new DNA into the nucleus that is to be put into the fertilized egg cell. Figure 10-1 shows such a technique. In this case, after fertilization takes place, we can insert new DNA directly into one or both pronuclei (one from the male and one from the female) using a microinjection technique (see Fig. 10-1). This allows the DNA of the fertilized egg cell to be transformed, though in a random fashion. Nonetheless, new DNA is inserted, and the fertilized egg can then go about becoming a whole organism.

The problem is that it is not certain that the DNA got into a place in the chromosome where it will be effectively used. The technique is rather random. Yet, once an animal has the new DNA inserted in an effective place, it is possible to breed that animal and get the particular trait into progeny as well.

Transgenic Mice—How to Make Them

Genetic engineering of animals has made important strides in recent years, especially in the development of transgenic mice. The approach incorporates many of the techniques we have previously discussed for bacterial and animal cells. We will outline the methods used with mice in some detail, realizing that such approaches can be applied to other animals as well.

The first step is to grow some totipotent cells in tissue culture. This is now readily done with early-stage blastomere cells, called **embryonic stem (ES) cells**. These cells are grown in culture medium just like the mammalian cell cultures discussed in Chapter 8. They are often derived from mice having a different coat color, which provides a marker that we will use later. An insert into the ES cells from black mice is made that contains the new DNA gene we wish to insert, a neomycin-resistant region, and a *lacZ* region. The ***lacZ* region** codes for an enzyme that cleaves a substance giving a blue stain. When the *lacZ* is present, the cells containing it are stained blue when X-gal (the stain substance) is present. Cells without *lacZ* will not be blue.

The new DNA can be delivered to the ES cells by microinjection, but is much more commonly placed in the growth medium and electroporation is then used to open the pores of the ES cells. The ES cells are then placed in a medium containing neomycin, which allows only the transfected cells containing the neomycin-resistant gene to grow. Those with the insert stain blue when X-gal is added (Fig. 10-2).

Using the latter approach, we can identify and isolate the transfected cells. It is also necessary to obtain blastocysts

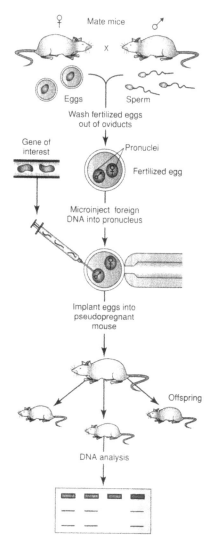

Figure 10-1 Overview of genetically engineering a mouse. Using fertilized eggs, new DNA is microinjected directly into one of the pronuclei in the fertilized egg, bringing new genetic material into the cell. Not all the DNA finds its way into useful places in the cell's chromosome. Using DNA analysis, it can be determined whether the DNA was inserted or not. Once the DNA is present and the appropriate trait is present, that mouse can be transferred to other mice through breeding.

from a white mouse cell line. These blastocysts contain numerous cells encapsulated in a membrane, as described earlier in this chapter. The transfected cells are then transferred to the blastocysts, after which the blastocysts

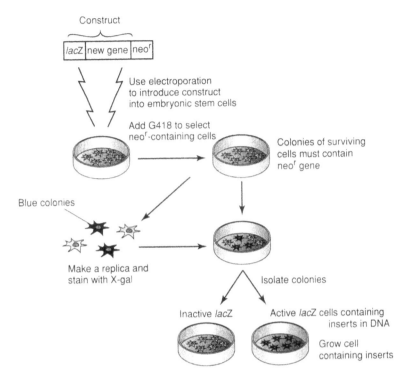

Figure 10-2   Transfection of a mammalian line. First, we sandwich the new gene between two markers (a construct), a neomycin-resistant gene (*neo'*) and a *lacZ* gene. By growing the cells on neomycin, those without the construct will not grow. We then select for those that received the entire construct by looking for blue colonies. When a cell is grown on a medium containing X-gal, the colonies that contain *lacZ* will turn blue.

are placed into the uterus of the surrogate mother mice (Fig. 10-3).

From these engineered blastocysts come chimeric mice. A **chimeric mouse** has new genetic information in some of its cells. By inserting the transfected cells into the blastocyst, those cells become part of the mouse to be born. It is never certain which portions of the embryonic mouse will be derived from the transduced cells. But if the new DNA is found in the germ-line (egg or sperm) cells, then the chimeric mouse can generate other chimeric mice as well.

The chimeric mice in the experiment outlined previously is generally black and white. The black-and-white coat color indicates that these mice contain the genetic inserts. By breeding

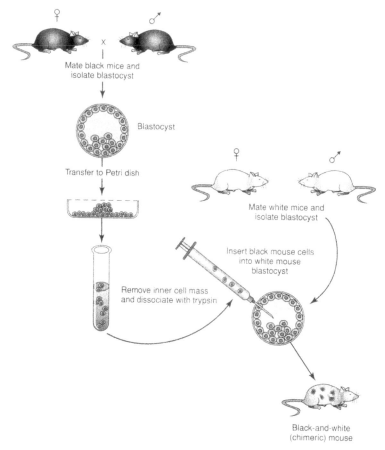

Figure 10-3 Overview of procedure for making a transgenic mouse. Cells from black mice are transfected as outlined in Figure 10-1. These transfected cells are added to the blastocyst obtained from a white mouse pair. The blastocyst is then inserted into the mother mouse again. The offspring are chimeric mice, containing both black-and-white coats.

these chimeric mice with other chimeric mice, entire strains of mice can be developed which contain certain new genetic characteristics. These are transgenic mice and can be very useful for research purposes in many ways.

For instance, some mice strains can be bred which are more susceptible to certain kinds of cancer. Perhaps the most famous of these was the so-called **oncomouse**. This mouse, produced at Harvard University in 1988, was highly prone to breast cancer.

The oncomouse could be used to test various cancer-causing agents and to test breast cancer therapies. Other mice have been developed with similar propensities for other diseases as well.

Knock-out Animals

One area of genetically engineered animals that has become extremely useful is in the development of mice strains with certain genes "knocked out." To make these, a faulty gene is incorporated into a mouse strain, as outlined above. The faulty gene is then transferred to the progeny and, through cross-breeding, the strain of mice with a faulty (knocked-out) gene is made. These knock-out strains provide important laboratory experimental animals, into which genetically transformed cells can be inserted to look for a resumption of the "knocked-out" function.

For instance, in one strain of mice, the entire immune system has been knocked out. This important finding has fostered additional experiments in which strains of mice have been developed which lack many genetic functions and even entire organs.

In other cases, certain genes have been knocked out and replaced by mutant human genes, giving rise to mice with certain genetic diseases, such as cystic fibrosis and Duchenne's muscular dystrophy.

In 1992, a strain of mice containing a "suicide gene" was developed. This gene promoted activity that destroyed the liver cells in the mice. Because the immune system is lacking, new human liver cells could be inserted into these mice, giving strains of mice capable of making human livers.

Clearly, the possibilities using the transgenic animal approach are almost limitless.

APPLICATIONS

Transgenic Animals—How They Are Used

Animals that had had new genes inserted into their germ line are called **transgenic animals**. Many such strains of animals have been and are still being developed. And many problems are yet to be solved in this arena. For instance, many transgenic animals are sterile, susceptible to diseases, and not suitable for further breeding. These problems stem from the difficulty of getting the new genes into exact positions in the cell genome. As noted in Chapter 8, several approaches are being developed to allow the insertion of new DNA into mammalian cells to occur at an exact site.

Certainly the ability to develop transgenic mice and other laboratory animals has dramatically increased the research possiblilties in disease-related areas. Many animals such as knockout mice are developed for research use only. Perhaps an even greater impact is being felt in the agricultural community.

Transgenic Farm Animals

One well-publicized improvement is milk production in cows. The pituitary gland in cows ordinarily secretes bovine somatotropin (BST), a hormone that induces cows to give milk. By injecting cows with BST, milk production can be increased dramatically (up to about 25%). BST has now been cloned in *Escherichia coli* bacteria and produced in commercial amounts. This has greatly facilitated the productivity in the dairy industry. But there is opposition and a general question as to whether such genetic engineering approaches should be used in food products (to be discussed more fully in Chapter 13).

Another area of active research is developing animal strains that are resistant to disease or infection particular to the animal species. For instance, chickens are susceptible to avian leukosis virus (ALV), a degenerative and often lethal disease. By inserting genes specific for the envelope of the virus into fertilized eggs, the chickens that developed showed significant resistance to ALV. Similar experiments with bovine leukemia virus (BLV) are being performed.

Some efforts have not worked well to date. Transgenic pigs, into which the gene for human growth hormone were placed, have not been robust. Often sterile and sickly, these animals seems to be much more sensitive to genetic alterations. On the other hand, inserts of sheep epidermal growth factor into sheep have made the shearing of the wool much easier.

Pharm Animals

Another entire line of experimentation is being carried out with farm animals. As noted earlier in Chapters 6 and 8, some proteins needed by humans cannot be readily produced by bacteria or perhaps even mammalian cell cultures. So, animals have been adapted to produce certain proteins and hormones used by humans. Such transgenic experiments have been made with cattle, pigs, sheep, and goats. The resulting animals are termed "pharm" animals—animals that produce human proteins in their milk or blood.

For instance, transgenic pigs can now produce human hemoglobin. A transgenic goat has been developed to deliver human tissue plasminogen activator (TPA) in its milk. Human lactoferrin is now produced by cattle. Many other experimental animals have been or are being developed using the versatile power of this approach.

Environment and Health

Insecticides are also being developed using genetic engineering. A recent article in *Science* trumpeted "Medfly Transformed—Official!" This announcement concluded over 10 years of research to transform the germ line of this pest. The medfly has cost the agricultural community billions of dollars and has cost the world hundreds of millions of dollars a year just to try to control its spread. The transformation occurred by injecting embryos of the medflies with suitable genes. Although the process is now in place, that actual transformation of the medfly to a nonpest is yet in the future. Nonetheless, the critical genetic engineering corner has now been turned.

Other experiments involving transgenic mosquitoes that will not carry malaria and snails that will not carry human parasites are under development. These approaches are the start of a radical new approach to solving some environmental and disease problems. The future is very bright in these areas.

CONCLUSIONS

It is possible to make transgenic animals that have new genes inserted into their chromosomes. This is done by genetically altering the early-stage cells and letting them grow into genetically altered whole animals. Research animals (notably mice) that lack certain genes can also be developed.

Such genetic transformation of human beings, though technically feasible, demands much more discussion on ethical issues before it is allowed. Such a discussion is clearly essential, because the promise of clearing certain familial lines of genetic disorders is important.

SUMMARY

Only by changing the genetic material in the nucleus of a fertilized egg will all the cells in an animal contain the genetic changes. At this time, it is not possible to selectively change portions of the DNA in a growing, fertilized egg cell. Nuclear implantation allows a new nucleus to be placed in a fertilized egg. This is technically difficult, but has been done in numerous cases, with the result that numerous identical animals can be clones.

By separating the early blastomere cells in an early-stage blastocyst, these cells can be grown in culture. Then, by using electroporation to allow new DNA to enter, new DNA can be inserted into some of the cultured cells. These cells can then be inserted into an encapsulated blastocyst, resulting in chimeric animals. In some cases, these animals have sperm or egg cells that have been modified with the genes as well. When these animals are bred with each other, entire transgenic animal lines can be developed.

Transgenic animals can be used to make new proteins, to grow new organs, and to be tested for medical uses. Development of these new transgenic animal strains provides unique opportunities for medical testing and production of new pharmaceuticals.

IV

HOW GENETIC ENGINEERING HELPS US

11

GENE THERAPY AND DISEASE

WHAT YOU WILL LEARN IN THIS CHAPTER

- How to diagnose genetic disorders using DNA probes
- How gene therapy is used to treat genetic disorders
- How to change defective somatic cells outside the human body
- How to change stem cells and other kinds of cells outside the human body
- How to target specific cells in the human body for genetic changes
- How genetic diagnostic methods and treatments are applied
- How genetic methods are used to treat cancer

The reason why we got into this whole discussion about genetic engineering was that we hoped it could help us medically. We have the tools needed to make changes in humans, but what kind of changes are warranted? It is now possible to screen a newborn baby or a fetus for genetic disorders using a number of different probes. The sequencing of the human genome, discussed in Chapter 12, seems to be providing more new genes every week. As new genes are found and other genes are shown to appear in damaged form, what can be done? First, let's discuss the diagnosis of genetic disorders.

GENETIC DISORDERS

Diagnosis

Many genetic diseases are already known. See Table 4-2 for a list of some common genetic disorders and their incidence. Clearly, many more could be mentioned.

It is now possible to determine who has genetic diseases using some of the genetic tools that we have learned. With **amniocentesis** or **chorionic villi sampling**, it is possible to withdraw fluid with fetal cells in it from the amniotic cavity of an expectant mother (Fig. 11-1). By extracting and disrupting the fetal cells, the DNA can be isolated. Then, it can be fragmented by

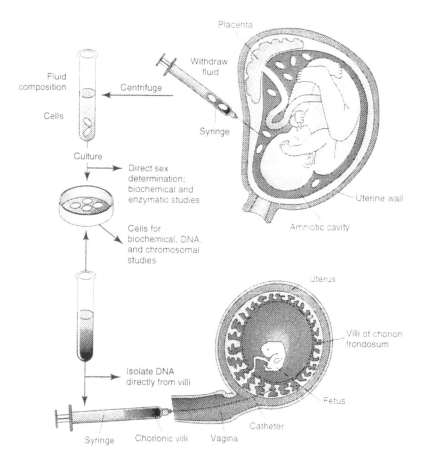

Figure 11-1   Amniocentesis and chorionic villus sampling techniques.

using a set of restriction endonucleases (enzymes). After this, the DNA can be run on an electrophoresis gel. As outlined in Chapter 6, the DNA is then blotted onto paper, and the fragments are screened using probes that are specific for certain genes.

The presence of an unusual gene signals problems ahead. For example, *BRCA1* and *BRCA2* genes have recently been found and sequenced. These genes are now thought to confer susceptibility to breast cancer in a small percentage of women. It should be possible to determine the presence of these breast cancer-susceptibility genes in an unborn child. In addition, normal genes with mutations, which might signal another genetic disorder, can be identified.

In practice, the analyses are seldom carried out using the Southern blot method outlined in Figure 11-2. Instead, the more common approach is to first attach a probe for a specific gene to a dipstick. The dipstick is dipped into the sample of disrupted fetal cells in which the DNA is denatured. In this step, hybridization (bonding) between the probe on the dipstick and any complementary regions of the fetal DNA take place. Then, the dipstick is rinsed and inserted into another solution in which a probe containing a fluorescent label is found. That probe is complementary to another portion of the target gene and will attach to it through hybridization. The dipstick is again rinsed and placed under ultraviolet light. If the gene is present in the fetal DNA, the dipstick will show it by fluorescing. This approach is called a hybridization **sandwich assay** and is still being developed for many different purposes (Fig. 11-3).

The sandwich assay can clearly be used to identify genetic disorders, but it can also be used to identify the presence of viruses and pathogens (eg, HIV). This approach and many variations of it are being developed rapidly by many companies to allow genetic screening of individuals to occur. Needless to say, many ethical problems surface in this approach, because genetic information of this nature would be useful to insurance vendors and others. We will discuss these and other ethical problems in Chapter 13.

Treatment

Once a genetic disorder is found, what can be done about it? Actually, it would be nice to fix the genetic disorder *before* conception so that it wouldn't exist. Although there are many available methods to do this, it would be enormously difficult to

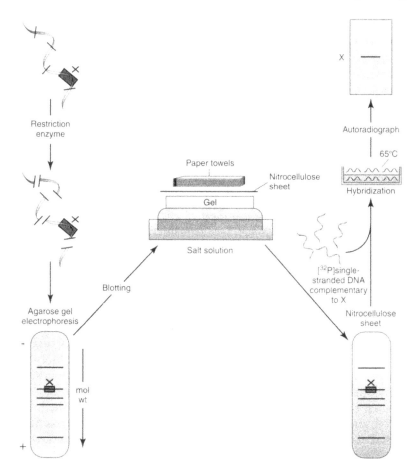

Figure 11-2   Identification of a specific gene (gene X) using restriction fragments and Southern blot technique. See also Figure 6-10.

accomplish. This is because sperm is extremely difficult to manipulate because of its size. The egg is much larger, but the problem of finding the correct gene, deleting the old gene, splicing in the right gene in the right place, and then using that egg to produce children would be extremely expensive and difficult, if possible at all.

The ethical considerations dealing with changing a single individual, and all of his or her progeny as well, have not been fully satisfied either. So, at this time, genetic manipulation of

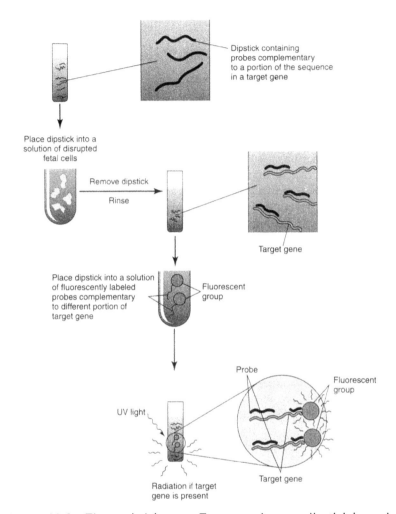

Dipstick containing probes complementary to a portion of the sequence in a target gene

Place dipstick into a solution of disrupted fetal cells

Remove dipstick

Rinse

Target gene

Place dipstick into a solution of fluorescently labeled probes complementary to different portion of target gene

Fluorescent group

Probe

Fluorescent group

UV light

Target gene

Radiation if target gene is present

Figure 11-3 The sandwich assay. For convenience, a dipstick is used, which has a short DNA probe bound to it that is complementary to a portion of the target gene. This dipstick is dipped into a solution of cells which have been disrupted so that their DNA is available to the probe on the dipstick. The dipstick is swirled for a short time, after which it is removed and washed. The dipstick is then placed in a solution that has another DNA probe, complementary to another portion of the target gene. To this probe is attached a small, fluorescent molecule. The dipstick is swirled, then removed, and rinsed off. If the fluorescent probe sticks to the target DNA, then by shining ultraviolet (UV) light on the dipstick, the dipstick will fluoresce. The presence of the fluorescence indicates that the target gene is present in the cell. This technique is being developed for an increasing number of genes.

human **germ cells** (eggs or sperm) is illegal.

Manipulation of Somatic Cells

The other approach to gene therapy is to treat somatic cells to repair genetic disorders. **Somatic cells** refer to all cells that are not germ (egg or sperm) cells. Altering somatic cells can change certain cell lines, such as blood cells, but it will not change the entire organism. Nor will such changes be transmitted to children. Genetic changes of this nature are allowed, but with strict regulation from the government.

Changes in somatic cells can be brought about using one of two major approaches: (1) gene manipulation of cells *outside* the human body or (2) specific targeting of regions *in* the human body, which may target a certain cell type, but ignore all others. We will discuss each of these.

GENE MANIPULATION OF CELLS OUTSIDE THE HUMAN BODY

The overall approach using gene manipulation is to remove cells with defective genes from a patient, genetically manipulate these cells to give them a correct gene, and return them to the patient. This approach has generally been used for blood cells because they are easily available and transferable.

Stem Cell Manipulation

All blood cells come from a single type of cell, the **stem cell**, found in the bone marrow. If we could transform some of these stem cells, it might be possible to fix some genetic disorders of the blood. So how do we go about doing this?

First, it is essential to isolate the stem cells. This has not been possible until recently, but now a few experiments have taken place in which stem cells have been modified. This approach has seen the most success in very young people, since their stem cells are quite active. Older patients have less active stem cells. However, it is expected that by using chemicals to stimulate the activity of stem cells, such therapy may also be available in the future for older patients. At present, there is no concrete evidence that any gene therapy involving stem cells is totally effective, although there is one clinical trial involving newborn infants in which some success has been observed.

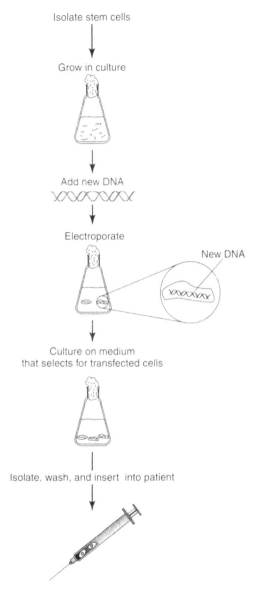

Figure 11-4 Stem cell manipulation. Stem cells are totipotent cells from which all other blood cells are made. By changing the DNA in the stem cells, we can change the DNA in all of the blood cells of an individual. Once the stem cells are isolated, new DNA can be inserted by electroporation, as we have outlined before (see Fig. 10-1). Transfected cells are identified and isolated and then grown to suitable number to be inserted back into the patient.

The experimental approach is to use the patient's own bone marrow, isolate the stem cells, genetically alter the stem cells, and grow them in culture (Fig. 11-4). When there are a sufficient number of stem cells—and large numbers are needed—the altered stem cells are reimplanted into the patient's own marrow.

Ultimately, the challenge lies in modifying the genetic information in the cells themselves. As noted in Chapter 10, we have tools for modifying mammalian cells in culture. The most promising approach is to use a crippled virus particle to deliver the modified gene to stem cells in the bone marrow. The methods of choice involve viruses, but targeting the DNA exactly to active portions of the genome remains a problem.

Manipulation of Other Cell Types

In addition to stem cells, techniques to manipulate other cells from the skin and membranes, as well as the precursors to muscle cells (myoblasts) are presently being developed. A number of gene therapy and gene transfer clinical trials are now ongoing, with many others awaiting approval. The crippled virus approach is the most actively used at this time, but is still often hampered by an inability to always insert the new genes into functionally active regions of the genome. This in turn results in marginal yields of the needed proteins from the genetically altered cells.

TARGETING OF SPECIFIC CELLS IN THE HUMAN BODY

Two different approaches are possible for targeting cells: the organ-targeting approach and the cell-targeting approach.

Organ-Targeting Approach

The organ-targeting approach has been developed to some degree, but still has room for improvement. In this procedure, vectors (carriers) bearing corrective genes are inserted directly into the tissue where they are needed. For example, in patients with cystic fibrosis, which impairs the lungs, vectors carrying corrective genes have been introduced directly into the lining of the bronchial tubes. This procedure has been partially successful. Similar approaches have been used for those with muscular dystrophy and cancer tumors.

The major problem that remains is targeting various organs in a safe and effective way. This is more pronounced when the therapy takes place in the body. Because genes are inserted randomly into cellular genomes, they might disrupt a tumor suppression gene, which would normally protect the body against cancer. Thus, the use of a **liposome carrier** (really just a lipid-bilayer structure to hold nucleic acids) or other nonspecific carrier, which tends to insert genes randomly in various organs, is not considered ideal for treatments of cells within the body.

Although crippled retroviral vectors can be successfully used in some cases, these vectors work only on cells that are actively dividing. Recent experiments have begun using crippled HIV as a vector to deliver genes to specific cells. This vector is able to insert new genes into nondividing cells, which makes it much more versatile as a prospective vector for cells that are not dividing, such as neurons. It has been shown to work in living organisms, giving it great promise. However, there is some reservation about using even a crippled HIV vector, since there is a chance that it might be able to spontaneously recombine with other genetic material and become virulent.

Cell-Targeting Approach

In the organ-targeting approach, a vector was developed and placed directly into the tissue to be genetically altered. Another approach, the cell-targeting approach, is to specifically target cells using vectors that will seek out and find only certain cells and deliver their nucleic acid only to those cells (Fig. 11-5). This technique is not yet being used for therapy, although good progress has been made in targeting specific vectors to specific cells. This is done by attaching specific markers on the outside of the vectors. These markers are recognized by receptors on the target cells.

The major problem with targeting specific cells is finding a way to allow the insertion of the genes into the target cell. At present, although the vectors seek and bind to target cells, there is no reliable way to induce them to insert their modified genes. The HIV vector discussed previously may provide an answer to this problem, since it can be packaged in a coat that would contain specific markers. However, much more work needs to be done before this is fully developed.

It is expected that the delivery problems will be solved in the not-too-distant future. Then, the approach of targeting specific cells within the living organism should become very powerful and offer tremendous potential in healing many types of genetic diseases.

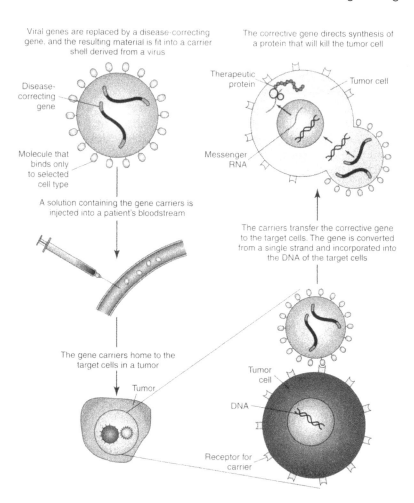

Figure 11-5   Targeted cell therapy. A virus that is specific for a certain cell line is modified by removing a portion of its DNA (see Fig. 8-6) and replacing it with a new gene that will kill the tumor cell. The altered virus particle is then inserted into the bloodstream and seeks out the cell line (a tumor) for which it is specific. It attaches and inserts new DNA into the tumor cell, making a protein that will kill the tumor cell.

APPLICATIONS OF DIAGNOSTIC AND TREATMENT APPROACHES

DNA Screening Tests

About a dozen companies are now investing significant amounts of money and time into developing diagnostics using the hybridization approach, in which certain genes are sought out and checked.

The dipstick method outlined earlier is becoming very useful for screening for genetic variations. It is especially useful for identifying defective genes in fetuses. Such tests are somewhat expensive and are not routine, but do provide a direct method of testing for potential genetic disorders.

The major problem looming in this area is not technical, but ethical. How much DNA testing should be done, when should it be done, and how public are the results? These and other questions are discussed in Chapter 13.

Gene Therapy

Table 11-1 summarizes the clinical trials on vectors of the National Institutes of Health (NIH) Recombinant DNA Advisory Committee (RAC). Over 100 trials are approved, but the approaches have both advantages and disadvantages.

The major problem still lies with the vectors. The most popular vector is the retroviral vector. A crippled version of a mouse retrovirus has been loaded with therapeutic genes and

TABLE 11.1 Vectors in RAC-Approved Clinical Trials

Vector	No. of Clinical Trials	Pluses	Minuses
Viral			
Retrovirus	76	Efficient to transfer Easy to make	Small capacity Random DNA insertion Dividing by cells only Replication risk
Adenovirus	15	Nondividing cells Possibly targetable	Immunogenic Replication risk
Adeno-associated virus	1	Nonimmunogenic	Small capacity Hard to make
Herpesvirus	0	Nonimmunogenic	Risks unclear Hard to make
Nonviral			
Lipsosomes	12	No replication Nonimmunogenic	Low efficiency
Naked or particle-mediated DNA	3	No replication risk Nonimmunogenic	Low targetability Low efficiency

used in 76 of the 106 trials approved to this date. Although this is the most efficient agent yet, the rates of transfer and expression vary enormously. The retrovirus is limited for use in rapidly dividing cells, so it will not target cystic fibrosis where the target cells don't divide. The major drawback is, as noted previously, that retroviruses insert their DNA randomly into host DNA, posing a possible threat of cancer.

Adenoviruses have also been used with some success. These viruses are DNA viruses that can cause colds and conjunctivitis in humans. However, they have been successfully harnessed, in crippled form, to be used as vectors to transmit new DNA into mammalian cells. Adenoviral therapy has problems in transferring sufficient quantities of genes into patient's cells and apparently causes an immune response, making multiple applications less likely. This approach has been used mainly in the effort to transfer the gene for cystic fibrosis (CFTR) directly into the bronchi or lungs of patients.

Case Examples

In the summer of 1990, a research team from the NIH received permission to attempt gene therapy on two girls, one age 4 and the other age 9 years. These children were born with a disease of the immune system called **severe combined immunodeficiency disease** (SCID). This disease has been found to be caused by deficiency of the enzyme, adenosine deaminase (ADA). Without ADA, toxic chemicals accumulate in the body.

The approach involved introducing the ADA gene into lymphocytes (white blood cells) removed from the patient. This was to be done by extracting the cells and exposing them to billions of crippled retroviruses carrying the necessary gene. The cells would then be reintroduced into the patients. The experiments were outlined in great detail, and permission was sought and granted from the necessary state and federal regulating agencies. After treatment, the girls improved immediately, and soon they produced the required level of the ADA to allow them to function almost normally.

In these cases, the government required that, along with the gene therapy for the girls, a standard treatment be given in which polyethylene glycol-ADA (PEG-ADA) was administered. The ADA enzyme is isolated from cattle and attached to PEG molecules, which prolong the activity of the ADA in the body. This provides a short-term boost in the ADA, preventing for a time the buildup of the toxic chemicals.

PEG-ADA treatment had been previously approved by the Food and Drug Administration (FDA). After 3 years of treatment, more than 50% of the circulating T cells in one patient contained the new gene, and only 1% contained the T cells in the other patient. Genetic therapy may have been working in one patient, but it is a not completely proven therapy.

In 1993, a similar effort was made with three ADA-deficient baby boys at birth. The aim was to target their stem cells by using umbilical cord blood injected with the ADA gene. Up to 10% of their circulating T cells now carry the healthy gene. The hope is that this growth will continue with time. Nonetheless, with these three boys, PEG-ADA is also being given. This is done because the physician felt it would be unethical to withhold it, since PEG-ADA was the previous FDA-approved approach. However, the amount of PEG-ADA is being reduced because it was found that PEG-ADA keeps genetically incompetent cells alive and counters the effects of the gene therapy to a degree. It is hoped that the PEG-ADA treatment can be eliminated entirely.

If the transplanted genes ultimately sustain these three patients without the PEG-ADA treatment, it would be the first solid demonstration of gene therapy curing a disease.

An effort in March 1996 was carried out to transfer a healthy gene to cure two young girls with Canavan's disease, a progressive illness that destroys the myelin sheath of nerves in the brain. The vector in this case consisted of a human gene (for aspartoacylase, the missing enzyme) coupled to an adenoviral plasmid (to insert the gene into human DNA accurately) and a liposome-polymer carrier to carry the genetic material to the deficient cells. The material was injected directly into the brains of the two girls. Although this procedure was carried out in New Zealand, the researchers did most of the research in the United States before being transferred to New Zealand. All the required paperwork to receive approval in both countries was done. Time will grade the performance of this approach.

The Promise of Gene Therapy

In spite of the relative lack of reportable success at this time, many clinical trials are being scheduled using gene therapy for genetic diseases, and many more will be forthcoming. In most cases, gene therapy offers the only possible permanent solution to their ailment. The shortfalls of some of the techniques will be remedied, and new techniques will provide additional strength to the procedures. There is little question that genetic therapy will become an increasingly active arena in treating genetically induced disease.

Cancer Treatment

Tumor-Infiltrating Lymphocyte—A Magic Bullet?

There are many types of cancer, so a single approach will not solve all cancer-related problems. In tumor-producing cancers, a lymphocyte called **tumor-infiltrating lymphocyte** (TIL) has been used as a weapon against such tumors. TIL is modified with a retrovirus vector containing a gene for the tumor necrosis factor (TNF), which destroys cancer cells. The TNF-producing TILs are then put into the patient with the tumor and seek out the tumor, releasing the TNF in the patients. Similar approaches are being developed for other kinds of cancer.

This "magic bullet" approach bodes well, not only for cancers, but other organ-specific genetic disorders. The major problem remains that the targeting is not always exact between the carrier and the target cell, and the placement of the DNA in the targeted cell is random, owing to the retrovirus vector used. These problems are expected to be markedly improved in the near future.

A PERSONAL GLIMPSE

A diary entry on August 25, 1989, of an 8-year-old girl with cystic fibrosis (CF) reads: "Today is the best day ever in my life. They found the Jean for Cistikfibrosis."

This "jean" had been looked for over many years. In 1978, two researchers found the first genetically linked gene—one that traveled with the sickle cell disease. This finding led the way for the study of genes to diagnose disease. In 1982, scientists began looking for such linkages between gene markers and CF. Massive screening was carried out, using a large number of families who showed the inherited trait. Combining this process with a company that had developed over 200 new markers, further screening showed that one of the markers was linked to CF.

In 1985, other researchers found another close marker on gene 7, but didn't know whether the CF gene was flanked by these markers or was just nearby. finally, in 1986, it was shown that the marker flanked the CF gene, and the race was on. The two markers were still 1.6 million basepairs (bp) apart, so there was room for hundreds of genes.

Using overlapping fragments to determine the sequence would have taken 18 years (at that time) to accomplish. But with some luck and lots of hard work, researchers were able to further isolate the area of the gene in about $1^{1}/_{2}$ years. Using cDNAs from chickens, mice, and cows, they were able to locate the CF gene region and then the gene, made up of 27 fragments, between which were intron sequences. The CF gene was made of 250 kb (1000 bp), and patients with CF lacked 3 bp. This deleted one of the 1480 amino acids in the protein for which the gene coded. This work was published in *Science* on September 8, 1989. For the 1 in 2000 children born each year with CF, there is now a greater hope for the future.

CONCLUSIONS

Gene therapy has a promising future in helping mankind with various genetic disorders. Clearly, it has many possibilities for diagnosing as well treating genetic disorders of the blood. With other disorders, finding ways to insert new genes into the specific types of cells remains a challenging problem. Nonetheless, increasing numbers of gene therapy treatments have been approved on a trial basis because there are just not good alternatives.

SUMMARY

It is already possible to screen DNA to determine the presence of certain genetic disorders. Although universal screening presents some ethical problems, screening for individuals with disorders or for fetuses with potential disorders is relatively straightforward. Dipstick assays are already in place to screen for an increasing number of genes.

Treatment of genetic disorders remains the challenge. Some genetic disorders, especially those of the blood, appear to be very possible to treat. By changing the genetic makeup of the stem cells, which makes all other blood cells, and placing these back in the bone marrow, genetic changes can take place permanently.

It is also possible to target certain types of cells for a genetic alteration. These methods are still developing. One kind of approach involves removing cells from the impaired person, genetically altering these cells, and returning them to the donor. Another kind of approach targets specific cell types within the body. In all cases, the challenge is still to get the new DNA into the right place in the cell genome, so that correct proteins or enzymes can be made.

12

OTHER
APPLICATIONS FOR
GENE THERAPY

WHAT YOU WILL LEARN IN THIS CHAPTER

- How to map and sequence the human chromosome
- How to identify alleles on chromosomes
- How to use restriction fragment length polymorphisms (RFLPs) to identify individuals
- What a variable number tandem repeat (VNTR) is and how it can be used to identify individuals
- What polymerase chain reaction (PCR) is and how it can make numerous identical copies of a strand of DNA
- How PCR can be used to amplify short tandem repeat (STR) units to identify individuals in forensic work

Up to now we have talked about how to put new genetic material in correct places in a chromosome. But we have really skirted the issue of how we can *locate* genes in human cells. Now we need to face it squarely.

ABOUT THE CHROMOSOME

As previously discussed, the genetic information in a chromosome consists of a long, uninterrupted strand of DNA, which contains many genes. In higher-order cells, each chromosome can contain many thousands of genes. In humans, there are 46 different chromosomes joined in 23 pairs. The amount of information contained in these chromosomes is immense. To work with these chromosomes, it is essential to map them, somewhat like a road map, by showing the relative positions of the various genes or other interesting sites.

In the last few years, there has been a worldwide effort to obtain the complete sequences of chromosomes of bacteria, yeast, mice, and humans. The tremendous progress in this arena predicts that the time is rapidly approaching when complete sequences of some representative species will be available.

Although the sequencing methods are straightforward in principle, sequencing such large amounts of DNA, as in humans, involves a vast amount of work. The shortest human genome is 50,000 kb long. This means that about 4500 fragments of DNA in this one chromosome have to be sequenced and linked. The task is enormous! Yet, it has now been accomplished to some degree.

The target date for the human chromosome to be completely sequenced is the year 2005, but it is increasingly apparent that this target may be hard to reach. By that time, the genome sequence will be known within an accuracy rate of 99.99%. This is an amazing feat that promises significant hope for prevention of genetic disease in the future.

Already, while in the process of mapping the genome, scientists have discovered thousands of new genes, some that lead to genetic susceptibilities to breast cancer, lung cancer, aging, and participation in high-risk sports. In 1990, only a handful of genes were identified for various genetic disorders. Now over 5000 have been identified, with new discoveries almost daily. These discoveries promise great potential for solving some of the critical genetic diseases in the future, because we will know the position and sequence of the genes involved.

Notable among recently discovered genes are *BRCA1* and *BRCA2*, which are found in familial-linked breast cancer patients. Although these genes are found in less than 10% of such patients, such a finding still provides a tremendous gain in the study of the effects of genetic transformation on this dreaded disease.

Another gene that was recently identified is the gene that causes Werner's syndrome, a rare genetic disease that causes premature aging. Investigators located this gene on the short arm of chromosome 8 using genetic linkage studies. The suspect region was sequenced—650,000 bases—and the mutated region was discovered. The protein that was found contains 1432 amino acids and apparently functions to unwind DNA in normal cells. The mechanism by which aging is accelerated when this protein is dysfunctional is not presently determined.

DNA PROFILING

There are many related applications and tools other than those used in genetic engineering that have developed as a result of efforts to identify differences in DNA. One of these is DNA profiling or typing (fingerprinting). The initial question was whether DNA from various organisms could be used to identify the relationship of one organism to another. By looking for organisms that are most closely alike, one might be able to determine the relationships among families of organisms and plants and other living species.

To better explain how these relationships are determined, we need to go back to the work of Gregor Mendel. As Mendel concluded his analysis of pea genetics in 1866, he recognized the gene as a "particulate factor," which passes unchanged from parent to progeny. He also recognized that a specific gene could exist in alternative forms (called **alleles**) that determine some particular characteristic, such as the color of a flower. When two chromosomes are paired together, the chromosome pairs exchange pieces in a process called **crossing over**, resulting in new unique chromosomes, which contain traits of both chromosomes (Fig. 12-1). These provide the genetic makeup of an individual.

It was postulated early, and has been confirmed since, that genes that are close together will not be broken apart by crossing over as often as those that are more distant. This tendency to remain together is called **genetic linkage**. So, the closer genes lie together, the fewer recombinant events occur between them. Conversely, genetic linkage can be taken as a measure of physical distance between genes. By looking at this linkage, the ancestry of a given species can be determined with some degree of exactness.

RESTRICTION FRAGMENT LENGTH POLYMORPHISMS

Another approach to assessing genetic differences can be made by looking at patterns made by restriction fragments. As we have discussed before, when a chromosomal fragment is cleaved

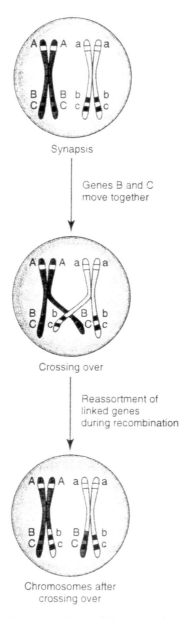

Synapsis

Genes B and C
move together

Crossing over

Reassortment of
linked genes
during recombination

Chromosomes after
crossing over

Figure 12.1   Crossing over. One of the several ways by which genes are exchanged in a cell. In this process, two adjacent chromosomes exchange genetic information by exchanging a portion of their genes. This allows for variation in the genetic makeup of individuals. Note that genes that are physically close to each other would have a greater tendency to stay together than those genes that are physically distant from one another. This is called genetic linkage.

by a certain restriction endonuclease (enzyme), fragments of specific sizes are made. These fragments are then run on electrophoresis gels and give a pattern of bands that depend on the length of the fragments. A particular piece of DNA always gives the same resulting fragments when a specific restriction enzyme is used. But if there has been a mutation that affects one of the restriction sites, an altered set of fragments will appear. This variation in the fragment pattern is called a **restriction fragment length polymorphism** (RFLP) (Fig. 12-2).

RFLPs occur frequently enough in the human genome to be used for genetic mapping. The RFLP approach can help us to discover whether certain people are related and to assess

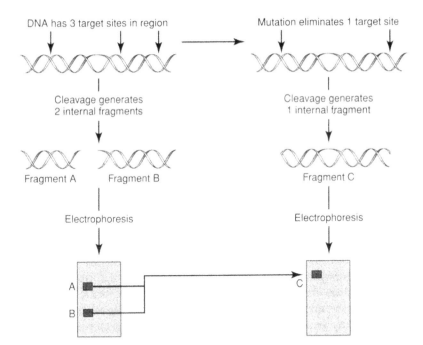

Figure 12.2 Restriction fragment length polymorphism (RFLP). See also Figures 11-2 and 6-11. Differences between individuals occur often in some segments of DNA. These are called polymorphic regions. Restriction enzyme digests of these regions produce various lengths of DNA. One of the ways by which this variation in length can occur is when a mutation takes place that removes a restriction site, as shown here. In these circumstances, only one band appears in the gel pattern after the mutation, whereas two appeared originally. Such differences in various regions of the DNA help identify certain individuals.

genetically linked characteristics. The main benefit of the RFLP approach in genetic engineering is to narrow the search for a particular gene to a defined region of the chromosome. If a particular trait is known to be inherited with a known genetic marker, then it is necessary to obtain only the portion of DNA with that marker in it and to perform the further assays to identify the gene. The RFLP map helps provide information on the place to look.

RFLP analysis is also commonly used to identify familial relationships in animals and plants. For instance, RFLP testing of families of wolves will indicate parentage and identify particular packs. Similar testing with mountain gorillas, various bird species, fish species, and many others has allowed geneticists to identify families or species containing certain traits.

Another use for DNA profiling is as a tool for "sleuthing," that is, to help ascertain whether a specific individual was involved in a crime of some nature. Ideally, to match individuals to one another or to a tissue sample, we should sequence their entire genome and look for variations. But within the human genome, there are over 3 billion nucleotides. This makes routine sequencing of human DNA impossible as an analytical tool. So two approaches—RFLP analysis and polymerase chain reaction analysis—are presently used, with many variations on each, to allow suitable comparisons to be made.

RFLP Analysis

Most regions of the human genome vary little from one individual to another. But some regions, which apparently are not used for structural or functional needs, vary greatly from person to person. These regions are **polymorphic** (having many forms) and are often unique to an individual. We can isolate and characterize such polymorphic regions and then use them to identify an individual much more conveniently than we can sequence the entire human genome. But we have to remember that the region being analyzed is not entirely unique and that there may be one or more individuals who have identical polymorphic regions.

Some polymorphic regions have recently been found to contain multiple identical sequences, which range from a few to about 60 nucleotides in length. The number of such repeating units varies substantially from individual to individual. At each end of these repeating units are identical flanking regions. So when a restriction enzyme is used on these flanking regions, the length of the restriction fragment varies, depending on the number of repeating units present. This region of repeating, identical sequences is called a **variable number tandem repeat (VNTR)** (Fig. 12-3).

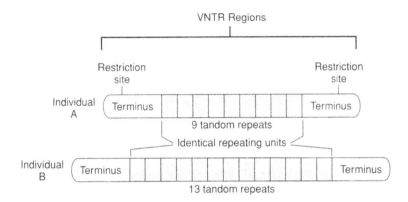

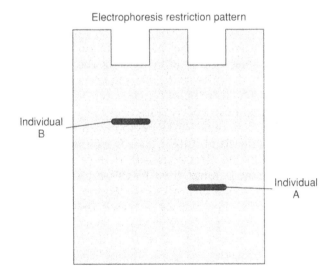

Figure 12.3 Variable number tandem repeat (VNTR) regions. In some polymorphic regions of DNA, segments are found with identical flanking regions (termini), but variable numbers of identical sequences that repeat over and over. All cells from a single individual contain a set number of these repeating units. But the same region in another individual may contain a different number of repeating units. The diagram shows an example of two individuals, one of whom has 9 and the other 13 identical repeating units in a specific VNTR region. This region is excised from the total DNA using a specific restriction enzyme.

Each VNTR region isolated from a single individual contains two fragments—one from each parent. These fragments are alleles. In some cases, both fragments are identical in length, indicating that each parent donated a VNTR region having an identical number of repeat units. When a single restriction enzyme is used to cleave the DNA from a single individual, two bands usually appear on the RFLP pattern. This is called a **single-locus pattern**. A single-locus pattern from one individual usually differs from that obtained from another individual unless the individuals are identical twins. To further establish the difference between individuals, additional VNTR regions can be used. When more than one locus is used, the analysis is called **multilocus analysis.**

The Procedure

So the approach to RFLP analysis is as follows. Initially, it is necessary to identify one of the highly variable regions in the genome. Using a specific restriction enzyme, one can cleave out (separate) the region containing this VNTR region, run it on an electrophoresis gel, and then blot it over to nitrocellulose paper. Using the labeled DNA probe that is complementary to the repeating sequence in the VNTR, the position of the VNTR-containing bands can be identified on the gel. The position of each band on the nitrocellulose paper is proportional to the length of the VNTR, which in turn is determined by the number of repeat units in the VNTR (Fig. 12-4.)

If we analyze a single VNTR region from two different individuals, two fragments of different sizes would ordinarily occur for each individual, as already noted. If the individuals were children of the same parents, their RFLP patterns could be identical or may differ in one or both of the bands. If a child were compared with his or her mother or father, one band should be the same and the other different, since one allele came from one parent and the other allele came from the other parent.

Identification by RFLP

One identical RFLP pattern is not sufficient to identify a specific individual, but if four or five such VNTR regions (loci) are used, the probability that the fragment lengths in all regions are identical becomes vanishingly small. Thus, a perfect match of the RFLP patterns at four or five VNTR regions provides exceptionally strong evidence that the DNA samples came from the same source.

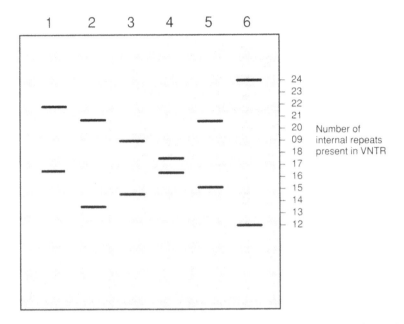

Figure 12.4 Restriction fragment length polymorphism (RFLP) pattern from six different individuals (lanes 1–6). The number of internal repeats is indicated on the right-hand side of the diagram. Note that each individual has two different lengths, one coming from each of the two alleles (one from each parent). All of these repeats come from a single variable number tandem repeat (VNTR) region. If similar patterns were analyzed from several VNTR regions, the combined patterns for a single individual would differ from each other considerably, even though an occasional fragment would have the same length, as shown here for one of the fragments in lanes 2 and 5.

These RFLP patterns were dubbed early on as "DNA fingerprints" because they were unique to individuals. By comparing patterns from an unknown sample and various individuals, identification could be aided. For instance, RFLP patterns have been useful in identifying relatives. A possible family "DNA portrait" using four VNTR regions (four loci) is diagrammed in figure 12-5. Note that each child derives one band in each allele from the father and one from the mother. Identical twins have identical patterns.

In 1994, the remains of Russian Czar Nicholas II and his family members were identified by means of DNA profiling. This particular family was apparently murdered in the Bolshevik revolution, and there was considerable question of identity of the children.

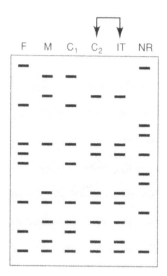

Figure 12.5   Restriction fragment length polymorphism (RFLP) patterns from family members. Four loci (four variable number tandem repeat [VNTR] regions) were used in this analysis. Each child (C_1 and C_2) has an individual pattern, with the bands being identical with either father (F) or mother (M) bands. Identical twins (IT and C_2) have identical patterns. A nonrelated individual (NR) has a pattern that doesn't match any of them, although there is one band of the same length.

RFLP analysis has also been very successful in determining the paternity of children. The patterns are readily used to identify fathers, since half of the child's genetic pattern should match that of the father (Fig. 12-6).

Another growing use for RFLP analysis is the diagnosis of genetic disorders. As increasing numbers of genes for genetic disorders are being identified and sequenced, it is possible in many cases to develop an RFLP analysis that will indicate the presence of the genetic disorder. The method by which this is done is outlined in Figure 12-7. In this case, a restriction enzyme that cleaves the DNA in the region of the disabled gene is used, thereby discriminating between a normal individual and a diseased individual.

Forensic Profiling

Although RFLP patterns can readily exclude relationships between individuals, identifying specific individuals exclusively is much more difficult. In forensic DNA work, such analysis can

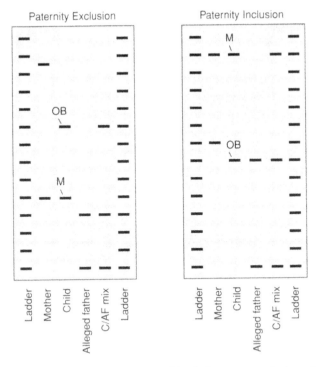

Figure 12.6 Paternity testing. Single-locus restriction fragment length polymorphism (RFLP) patterns show the patterns provided by mother, child, and alleged father. In the paternity inclusion pattern, the obligate band (OB) matches that of the alleged father. The mixture of the child and alleged father (C/AF mix) shows the two bands from the father and one from the mother and no others, providing strong evidence of paternity. In the paternity exclusion pattern, the obligate band (OB) does not match that of the alleged father, and the C/AF mix lane shows four bands, not three, providing strong evidence that the alleged father is not the real father. Additional loci could be tested to provide additional evidence.

yield particularly compelling evidence for such a match, but there is always a limited probability that two individuals may have identical RLFP patterns. Therefore, RFLP pattern matches must always be couched in probability of match. The more polymorphic positions on the DNA that are analyzed, the more certain the match.

Although such probability arguments have been made in the courtroom in the past and some DNA evidence was not used as the result, today DNA RFLP pattern evidence is generally

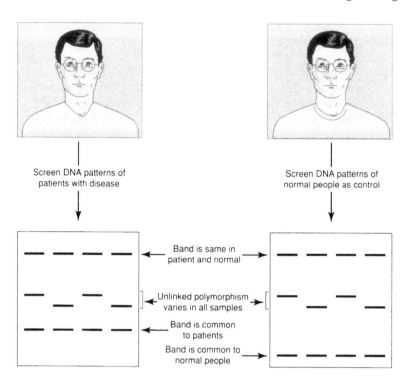

Figure 12.7 Disease diagnosis. Some genetic disorders can be identified using restriction fragment length polymorphism (RFLP) analysis. By identifying the region of the chromosome in which a genetic disorder may occur, analysis of that region using RFLP can provide evidence for the disorder. In the illustration, a single locus is used to identify patients with a genetic disorder. Note that in such patients, certain bands are identical with those obtained in normal individuals. However, a band or bands may be common to individuals with the genetic disorder and may appear at a different position than that which would appear in a pattern from a normal person. This kind of DNA analysis is especially useful for infants and unborn fetuses.

accepted. This approach was markedly enhanced by the O.J. Simpson trial, in which the DNA evidence was allowed, expertly presented, and shown to provide unequivocal matches on the blood samples.

RFLP analysis can also be used to identify pathogens of unknown origin. For instance, a mysterious disease started afflicting people in the "four corners" area of Utah, Colorado, Arizona, and New Mexico. With RFLP analysis techniques, the disease vector was identified as a hantavirus, and appropriate steps were taken to diagnose and treat the illness.

Polymerase Chain Reaction—The Analysis

Polymerase chain reaction (PCR) is a technique that has been developed in recent years, which allows scientists to make large amounts of DNA identical with an original piece. For instance, if a very small amount of a bodily fluid or cells is present, PCR allows the DNA in those samples to be multiplied many times over, providing sufficient amounts to analyze. This approach has become especially useful in forensic work and promises to be the method of choice in all cases in the future.

The Procedure

First, a small amount of DNA is needed. Then, through sequence analysis, the sequence of the first 15 to 20 nucleotides at both ends are determined. Next, short pieces of DNA complementary to these regions are synthesized in great quantity (readily done with the chemical synthesizers commercially available). These are called **primers.**

Then, by adding a heat-resistant DNA polymerase (*Taq* polymerase) under carefully controlled conditions, new strands of DNA are made, which are identical to the region found between the two end regions to which the DNA primers were bound.

For example, in the usual mode of operation, the DNA, primers, and *Taq* polymerase are placed in a vial and heated to 94° C for 1 minute to denature the DNA (dissociate the two strands). The temperature is lowered to 55° C for 30 seconds to allow the two primers to anneal (bind to the end regions to which they are complementary) to the ends of the each strand of the target DNA. Then, the temperature is raised to 72° C for 1 minute, during which time *Taq* polymerase makes two new strands of DNA, each complementary to one of the original parent strands. The sample is heated again to denature (separate the strands of) the newly formed DNA, and the cycle is repeated. Each time all of the newly made strands become templates for the next round of synthesis, resulting in over a million copies of the original DNA in a matter of hours using 30 to 35 cycles.

PCR is both rapid and convenient (Fig. 12-8). But *Taq* polymerase has no proofreading mechanism, so there is a greater chance for error in replication than in living systems. Statistically, the error rate is about 1 mistake of every 10,000 nucleotides synthesized. Although this error rate is still low,

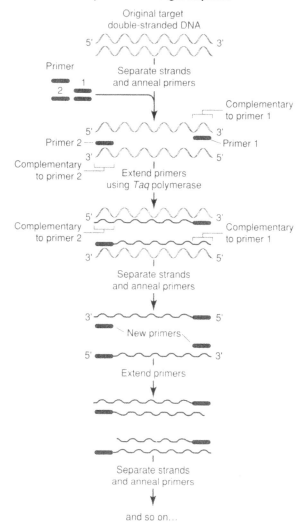

Amplification of Target Sequence

Figure 12.8 Polymerase chain reaction (PCR). This technique allows a large amount of DNA to be produced which is identical with the original DNA sample. The initial DNA is separated by heating, and primers that are complementary to either end are added and allowed to anneal (bind) to their complementary ends. DNA polymerase is added (*Taq* polymerase; it is heat-resistant), and this manufactures new DNA strands complementary to the original strands. The strands are then separated by heating, additional primers are annealed, and DNA synthesis occurs again. This cycle is continued until enough DNA is made for analysis. See text for details.

mistakes made in the early periods of replication are amplified in the entire product, which may cause faulty information to be generated.

The great advantage of PCR is that even a single piece of DNA from a single cell can be used as a test sample and will give sufficient product for further analysis. In addition, it is possible to use fragments of DNA from samples that may have been damaged as templates, allowing analysis in situations that would otherwise be impossible. A disadvantage of the PCR technique is that the longest DNA that can be amplified is no longer than about 5000 bp. This limits the size of possible regions of a single DNA that can be analyzed. RFLP analysis, on the other hand, is not so limited.

To make the PCR approach useful in DNA profiling, it is essential to identify a specific region of DNA and amplify this region. This region must be polymorphic (highly variable), yet have regions around it (flanking regions) that remain constant. Primers that are complementary to these constant flanking regions can be made and used in the PCR experiment.

It would be nice to use the VNTR regions that were used for RFLP analysis. The problem is that VNTR regions are too large to be amplified by the PCR technique. Recently, a number of **short tandem repeat** (STR) regions have been discovered, which contain repeating regions of 3 to 7 bp each. Although these are repeated multiple times, the aggregate size is easily accommodated by PCR amplification techniques. These regions can be amplified by PCR and then analyzed using essentially the same protocol as used with RFLP analysis. Multiple STR sites are now being used, which provide abundant information for identification purposes. The PCR-STR approach for identification is now the method of choice for almost all commercial laboratories, for the FBI, and for most state crime laboratories.

Clearly, the power of hybridization of nucleic acids with probes to specific regions allows an elegant, rapid, and dependable technique to identify almost unique characteristics of the individual from which the samples emanated.

SUMMARY

Mapping and sequencing the DNA in human chromosomes is a gigantic project, but is already paying dividends as new genes are being discovered. Diagnostic techniques are present and will continue to be developed. This will allow faulty genes to be identified. By identifying the positions of these genes, the possibility of genetically engineering cells to help alleviate some human genetic disorders becomes more of a reality.

In addition, the use of DNA as a "fingerprint" for identification purposes has blossomed tremendously. Restriction fragments can be used to identify family members readily. Such techniques are used not only for human families, but to identify relationships among animals and plants. RFLP analysis has been used in forensic work as well to help identify short tandem repeat (STR) regions, a number of which have been discovered. Using this PCR-STR approach, the donor of the sample tested can be identified with considerable certainty.

13

BIOTECHNOLOGY, SAFETY, AND THE FUTURE

WHAT YOU WILL LEARN IN THIS CHAPTER

- Biotechnology—what it is and what is isn't
- How a patent affects biotechnology industries and universities
- How safe is genetic engineering
- What safeguards are in place to ensure the quality of genetically engineered products
- What the human genome project is
- Ethical considerations in genetic engineering
- How genetic engineering may affect the future
- The future of biotechnology

All that we have learned in this book is of little use to mankind unless it can be applied and used. Along the way, we have mentioned various applications, to which some of the techniques have been applied. But the effort is much larger than we have room to discuss. Ever since the mid-1970s, when the original recombinant DNA work was made public, imaginative minds and billions of dollars have been spent trying to bring to pass the dream of manipulating genes for the good of mankind.

Commercial application of genetic engineering techniques falls into the broad category of biotechnology. **Biotechnology** is defined as "any technique that uses living organisms or substances from those organisms to make or modify a product and perform services, or to develop microorganisms for specific uses" (United States Office of Science and Technology). In a very broad sense, the manufacture of soy sauce, brewing alcoholic beverages, and bread and bagel baking all could be included in this definition. However, generally when we think of biotechnology, we think of processes that use genetic methods to increment the usefulness of natural organisms or to develop modified organisms in some way.

TECHNOLOGY TRANSFER

Patents

For the ideas and techniques that we have discussed in this book to make their way to the commercial market, a mechanism needs to be put in place to protect the inventor. Patents have been used for protection of inventors ever since the ratification of the Constitution. A **patent** is a legal monopoly on an idea or discovery, which provides the inventor with the right to use and market the discovery for 20 years to the exclusion of all others. The major problem in the area of genetic engineering was that it had long been held that living things were not patentable.

In the late 1970s, applications for various applications of genetic engineering were received and dismissed by the United States Patent and Trademark Office. The issue was litigated, and, in 1980, the United States Supreme Court was brought into the foray when Dr. Charkrabarty tried to patent a genetically altered oil-eating *Pseudomonas*. The question was whether it was naturally occurring. The Court ruled that it was not and therefore was patentable. This was the first time a living thing had been patented. The ruling was broadened in time, finally allowing genetically altered higher-order living organisms to be patented as well.

Since that time, patent rules, as they applied to altered living organisms (transgenic animals) and to discovery (and modifications) of portions of the human genome, have been refined. The following are major tests for patentability of a product or process:

- It must have **novelty** (that there was no previous finding like it).

- It must not be **obvious** (not obvious to a person familiar with the area).
- It must have **utility** (real usefulness in its application or design).

But should everything be patented? It is clear that by using DNA recombinant techniques and applying the methods that we have discussed previously, many new, nonnatural products and organisms can emanate. Recently, the National Institutes of Health (NIH) applied for patents for thousands of cDNA fragments being used in the human genome mapping project. This caused no small stir worldwide, because it would, in essence, patent the human gene. Happily, the NIH withdrew their application. The Patent and Trademark Office has since issued a statement indicating that such cDNA fragments would not be patentable. This was overturned in court and now cDNA fragments, a few at a time, are patentable.

Major areas of the patent regulations dealing with DNA and genetic engineering are still undergoing changes. Nonetheless, to date, thousands of patents have been granted with tens of thousands more on the way. Even though this effort is massive, the payoffs are massive, too. Biotechnology is expected to be a $50 billion/year business enterprise in the United States at the turn of the century.

IMPACT ON UNIVERSITIES

The intense effort to do applied research and to commercialize the resulting products has had an immense impact on the university systems in the United States and increasingly throughout the world. Many universities have now established technology transfer offices, so that patentable ideas by university investigators can return some profit to the universities. Faculty members are under scrutiny to determine whether the work they are doing has commercial value and, if so, what share the university will take. An ethical problem with faculty members comes when they must decide whether the research they are doing should be in the private sector or not or whether they themselves should be in the private sector as well.

So, in recent years, alliances have been formed among universities, scientific researchers, and biotechnology companies. To be successful, this alliance must be beneficial to all parties, so that exciting science performed in the university can be developed into marketable products. Getting the right mix is often difficult. There are many successful examples and many more failures.

Although some of these partnerships between industry and the universities are stimulating increased research at universities, some feel that the prime purpose of universities—that of acquiring and disseminating knowledge—is being eroded. Researchers who struggle to maintain basic research are finding it increasingly difficult to remain competitive as more money is directed to applied research. In addition, colleagues and sometimes administrators make it clear that applied research interests would be more important to the university.

Finally, note that there is no small financial incentive to pursue the route of applied research. A really successful idea can make millionaires of the inventors. These numbers make university professors dizzy. With such financial incentives, scientific investigation will increase, which in turn will spin off into new businesses and other commercial endeavors.

SAFETY

Two-edged swords cut both ways, and genetic engineering is a two-edged sword. Not only can we dream of the mighty effects on health and well-being that genetic engineering promises, but we can also dream about the possibility of some engineered organism going awry and becoming uncontrolled in the environment. Scientists have considered in depth the possibility of such a thing happening.

The Birth of Guidelines

When the power of DNA recombinant techniques (the insertion of new DNA into chromosomes) was first envisioned in the early 1970s, many scientists became concerned. To discuss these gargantuan social concerns and to discuss whether restrictions should be placed on recombinant DNA research, 139 outstanding scientists from 17 countries gathered at the Asilomar Conference Center near Monterey, California, in February 1975. For several days, scientists discussed their concerns and the prospects of recombinant techniques for the future. The discussion was intense and heated at times, with views being expressed on all sides of the issue.

The results of the Asilomar conference and later discussions were guidelines from the NIH. These came into being in 1976.

Shortly after, other national organizations, such as the Environmental Protection Agency (EPA) and the Department of Energy (DOE), also created guidelines. Initially, these guidelines were so restrictive that they essentially closed down some areas of research. Later, as new, disabled strains of bacteria and viruses were developed, the guidelines were relaxed and some of the restrictions were lifted, allowing research to move ahead.

Over time, scientists have become more and more aware of what may or may not present dangers either inside the laboratory or in the environment. The result is that, although there are still some stringent restrictions governing the use and development of some types of genetically altered species, most of these restrictions presently apply to organisms that might be released into the environment.

Within the controlled conditions of the laboratory, most experiments are allowed, with only a few demanding the totally **restrictive environment** (a totally contained laboratory in which a person must completely change clothes on entering and also on leaving), which was once thought necessary for all types of recombinant DNA work. Most scientists view the risk of an outbreak of disease or an unmanageable organism as being very small. A strong argument for loosening government restrictions is that Mother Nature has been doing genetic recombinations for millennia, and our efforts will likely do very little additional damage.

Lingering Concern

Some people are concerned about allowing genetically altered organisms into the environment. Present guidelines allow such a release only when it is proved that the organism will be completely safe. The safety of the organisms deals with the results of an accidental spill, accidental ingesting, or possible alterations that would make it a noncontrolled species in the environment. Although no guidelines are foolproof, the NIH guidelines are sufficiently strict to ensure a wide margin of safety.

The NIH has established a Recombinant DNA Advisory Committee (RAC). This group must approve all gene therapy experiments and experiments that put genetically altered organisms into the environment and must set the standards for laboratory safety in the handling of certain organisms. In addition, each university must have a local advisory committee, which likewise must give permission for experiments involving recombinant DNA. These safeguards provide independent observers to identify potential pitfalls in experimental procedure or unsafe protocols that may be used.

Ethics

In addition to safety, the problem of ethics looms over the entire field of gene therapy. Are we changing the future by genetically engineering bacteria or other organisms? Is the benefit of trying to make things greater than the cost of the damage that may ensue? To what degree should we be allowed to genetically change a human being? Is it desirable to try to genetically alter human children? Can the data bank obtained from the human genome project be used to screen future generations? What is to prevent **eugenics** (destroying individuals of a genetically distinct type) from occurring? Can an insurance company have access to a person's genetic profile? If an infant is shown by means of genetic screening to be mentally or physically disabled, should the parents be informed? These and other questions are being asked from many quarters and generally have no right answer.

It is interesting to note that 5% of the budget for the human genome project has been allocated to determine the ethical, legal, and social implications of the project. Four areas of ethical, legal, and social concern have been identified to date:

1. Privacy of genetic information
2. Safe and effective introduction of genetic information in the clinical setting
3. Fairness in the use of genetic information
4. Professional and public education

Each of these latter four issues has many facets to be considered as we deal with genetic issues. We have discussed the safety of engineering of bacteria and other organisms. It should be reemphasized that in the process of maintaining life, much nondirected recombination of DNA takes place naturally. Especially when unusual environmental stresses come upon cells, the cells react by changing their genetic makeup to meet the challenges. This kind of adaptation or directed mutation is constantly taking place around us. Even with an extensive amount of directed genetic engineering, we are not likely to come close to producing the amount of alteration that takes place naturally. From this standpoint, the future is constantly being altered, but mainly by "natural," undirected means.

If we can alter bacteria and plants in such a way as to make life a little better for humans, we can certainly justify our involvement ethically in genetic alterations. However, as we move up the scale a bit, we start to run into problems. Testing with laboratory animals has recently come under the close scrutiny of animal rights' activists. Do we as human beings have the right to subjugate lesser species to our sometimes inhumane testing procedures? These questions are not yet answered completely. We clearly need to test some ideas and products on living beings *before* we use humans. But how much is essential and what guidelines should govern such testing are yet to be determined.

The ethics as to whether we should genetically change human beings themselves goes deep into the center of our cultural system. There are actually two levels of change that we should consider. (1) There is a change to individuals, using genetic engineering techniques to alter their genetic makeup, but in ways that will not affect their children. (2) There is the possibility of genetically engineering entire lines of humans, much like we have learned to do with transgenic animals.

Changing or Adding Genes to Individuals to Cure a Genetic Disease

We have shown that it is possible to change the genetic makeup of individuals, especially those with diseases of the blood. This can be done to a limited extent by altering the genetic makeup of the bone marrow stem cells. In these cases, only the individuals so treated are affected, not their offspring.

The major ethical concern is whether it is right and proper to devote the substantial resources necessary to develop such technology intensive processes for the benefit of just a few individuals. In other words, although the process of developing genetic engineering techniques to treat human beings is intellectually satisfying, is the payback to the public sufficient to merit the investment?

While this question can be raised for many areas of science, it is true that genetic engineering approaches, at best, will provide treatment only for relatively few of the sick and diseased throughout the world. As the human genome project was conceived and discussed at length, many arguments against the project pointed to this very issue—that the enormous cost would benefit very few. Congress made the decision to pursue that endeavor. Since Congress represents the

American publie, it can be argued that Congress may feel that genetic engineering of individuals would be appropriate as well.

Genetically Altering Germ (Sperm or Egg) Cells

What about fixing genetic problems for entire family lines, such as the hemophiliacs? Are we justified in making the changes necessary to cleanse certain lines of genetic disorders? This problem is not resolved and is not likely be resolved for some time. At present, a total moratorium exists on any genetic experimentation with human cell lines. Still, it is clear, from the transgenic animal success, that such an approach is feasible. However, both scientific and ethical questions remain.

From the scientific standpoint, we know that certain disorders are caused by single mutations in a single gene. Yet, we also know that proteins often act in a coordinate manner. fixing a problem in one place may create a problem elsewhere. For instance, in some transgenic animal experiments, notably in pigs, inserting a new gene often makes the transgenic animal sickly or sterile. Humans are considerably more complex, and the results are much more difficult to predict. Clearly, complete scientific understanding is still lacking.

The ethical side of genetic alteration of germ cells is even more complex. How much do we change the future by altering the present? Would a transgenic family line retain all of the other beneficial characteristics that were present before the engineering took place? Who will make the decisions as to when germ cell alterations are allowable? Is the cost to society as a whole worth the benefit to a particular family? These and many other similar questions remain to be resolved before such experimentation is allowed.

The moratorium on transgenic experiments with humans is correct until the issues have been fully discussed and settled. Even with the moratorium in place, genetic testing still has fundamental questions to be resolved. One of these areas is the possibility of using genetic testing as a basis for eugenics, which is to control the hereditary characteristics of humans or other species.

Eugenics

If genetic testing were to become mandatory, is it desirable to lawfully withhold reproduction rights from those who do note meet societal standards? Even at this time, programs of mandated abortion are being carried out in some parts of the world, based in part on the genetic testing of the offspring. Should this be allowed? If so, to what degree? In what way will society be able to enforce a ruling in this area. Science fiction stories often tell of societies tailored through genetic testing (and engineering) to breed only those who are most fit according to the standards of that society.

The possible abuse of genetic testing methods by unscrupulous persons remains a real problem. In the United States, constitutional safeguards will likely prevent significant abuse, but other nations and civilizations are not so governed. Therefore, it is essential to mandate safeguards internationally.

Genetic Screening

Genetic Screening and Privacy

What about genetic screening? Should an insurance company be allowed free access to a person's genetic profile? Does this fall under the physician-patient privilege granted by law? It can be easily seen that genetic screens can be used to exclude potential high-risk clients for insurance purposes. Is this proper?

An argument could be that no genetic testing should be allowed. Still, this is unfair to the offspring, since early diagnosis may allow early and possibly more successful treatment of a genetic disorder. This issue is not resolved, but lines must be drawn to allow genetic screening without a risk of becoming uninsured. Steps are being taken to restrict access by insurance companies, as well as to restrict the use of such information. The issue of just how private genetic information really is has yet to reach the Supreme Court.

WHAT OF THE FUTURE?

Genetic engineering, like computer chips and lasers in the past, is an area of tremendous potential and virtually unlimited possibilities. We know how to make genetic changes, but the understanding of all of the complex interrelationships between genes and life is yet limited. In addition, ethical and moral considerations may place limitations on all that might be done.

The Future of Genetic Engineering as a Science

Since DNA was first transferred by design into an organism and expressed as a protein, scientists have been quick to imagine the potential of the tool that was discovered. Early experiments were limited to bacteria and viruses, followed soon after by those on animals and plants.

As a diagnostic tool, screening genes is a powerful and accurate technique for diagnosing genetic disorders. This power will continue to grow and more and more genes are discovered. It is probable that within the next decade, genetic screening of individuals will become as routine as blood tests now are. There remains the issues of privacy and ethics to be resolved, but the technology needed to provide such tests will continue to improve.

It is interesting to note that in so doing, all three areas—genetic testing, computers, and lasers—will be brought into play. Efforts are presently underway to use a computer chip itself as the "dipstick" for genetic probes, with a laser used to identify the positive reactions.

In the area of diagnosis, we can surely predict significant progress in the future. However, repairing genetic diseases will require a longer period to develop. For genetic disorders of the blood, there is significant hope for substantial progress in the next decade. However, in the case of many of the other genetic disorders, gene therapy will be considerably more difficult, mostly because of the difficulty in targeting the new genes to the appropriate cells.

Gene therapy is being used increasingly as a tool in cases in which no other remedies exist. Experimental techniques are rapidly developing, with novel vectors being developed to help target the new genes. Understanding of how the gene works and is regulated is continuing to increase. So, the promise of effective gene therapy remains bright for the future.

The development of transgenic animals for research and pharmaceutical purposes is a rapidly growing and promising area. For scientific research, the continuing development of "knockout" mice lacking specific genes is a powerful tool for research. Other laboratory animals having genetic alterations that allow them to be used for research purposes are rapidly being developed as well. And the pharmaceutical industry is investing substantially into the arena of "pharm" animals, which are genetically altered to supply needed proteins for health-related purposes. Such developments will continue to accelerate into the foreseeable future.

In addition, rapid acceleration is occurring in the development of genetically engineered animals for food purposes, such as catfish with added growth hormones, chickens engineered for leaner meat, and shellfish engineered to maximize growth. These and more are presently in use or under development, with many more in the pipeline.

The same acceleration is even more true in the area of plants. Increasing numbers of crop plants are being developed using genetic engineering to provide disease resistance and herbicide resistance and to carry genes for more rapid growth. These genetically engineered plants show great promise for agriculture. Although it is necessary to go through regulation and licensing to disseminate genetically engineered organisms into the environment, these restrictions are designed to protect the potential consumers. In the long run, the future looks very bright for increasing the numbers and varieties of genetically engineered plants that will be used as food products.

Biotechnology's Future

The future for biotechnology companies looks especially bright. Although many companies have experienced financial growing pains, there is continued expectation that biotechnology will provide many solutions to various challenges in the future. Billions of dollars are being invested to develop new and better products. Those having medicinal applications are rigorously screened by the Food and Drug Administration before they are licensed for use.

Various Uses for Genetic Engineering

Genetically modified organisms will certainly be harnessed to many industrial uses, such as waste clean-up, harvesting otherwise unobtainable oil reserves, and other important tasks. In addition, the varied conditions under which organisms grow (freezing temperatures, extremely high heat, high pressures) will be used in providing stable enzymes and products that can function under variable conditions in research and agriculture. An example is the use of *Taq* polymerase (a high-temperature enzyme) used in the polymerase chain reaction (PCR) approach we discussed in Chapter 12.

Considerable progress has also been made in other areas of genetic manipulation. Animal breeding has been revolutionized by artificial insemination and surrogate mother breeding. These standard procedures will be used in developing additional genetically modified animals for research and breeding purposes.

Monoclonal antibodies have been developed using genetic engineering techniques. These antibodies are capable of discriminating a single specific site on a complex macromolecule and have shown themselves to be highly specific reagents in advanced diagnostic techniques. Using these amazing antibodies, we can literally "see" inside the human body with great clarity.

One of the more exciting areas in genetic engineering is that of nitrogen fixation. If various grain crops can be engineered to fix nitrogen (take nitrogen from the air and convert it to nitrogen products used by plants) out of the atmosphere to supply their needs in protein manufacture, expensive fertilizers may no longer be essential.

All the latter applications, as well as many that are yet undiscovered, will undoubtedly affect our future. In many cases they will make life a little better. Still, these techniques are not a panacea. We cannot expect genetic engineering to solve social ills or disease induced by malnutrition. We also must not expect genetic engineering to eliminate even the simplest genetic diseases for some time. Nonetheless, the promise is there and if recent history is any indication, great surprises will await us as we pursue the research necessary to provide answers to the problems ahead.

GLOSSARY

A

acrylamide material popularly used in gel electrophoresis.

agar a gelatin-like material used in gel electrophoresis; more porous than acrylamide.

allele a portion of a chromosome that codes for certain traits. There are generally two alleles for the same trait in each chromosome—one from the father and one from the mother. Portions of alleles are often mixed together in the offspring.

alpha helix a helical secondary structure commonly found in the amino acid chains of proteins.

amino acids the building blocks from which proteins are made. In natural proteins, 20 amino acids are used (see Appendix for a complete list and their structures).

amniocentesis the process in which fluid is removed from the placenta for analysis.

anneal the process in which two complementary strands of nucleic acids are allowed to associate and form hydrogen bonds between the strands.

anticodon the three nucleotides on the transfer RNA (tRNA), which are complementary to the three nucleotides on messenger RNA (mRNA) that code (a codon) for a particular amino acid.

archive the place where information is stored; in living things, normally DNA.

aromatic a molecular compound that consists of carbon and other atoms and forms a ring structure.

assay any kind of test to determine the presence or absence of something.

atom the fundamental unit of an element, composed of a nucleus containing protons and neutrons and surrounded by electrons.

B

β-galactosidase an enzyme that breaks down lactose into two sugar molecules—galactose and glucose.

β sheet a zigzag secondary structure commonly found in the amino acid chains of proteins.

bacterial lawn a Petri dish that has the agar covered completely with bacterial colonies.

bacteriophage a bacterial virus; many have a geometrically shaped head unit, a tail, and appendages with which they attach to bacteria.

bacterium a small, one-celled organism that contains all the necessary machinery for life.

base a cyclic structure containing nitrogen, carbon, and other elements. It is attached to a sugar molecule to make a nucleoside or nucleotide, the basic building block of nucleic acids.

base deletion a form of mutation in which a normally occurring nucleotide is deleted from a gene.

base insertion a form of mutation in which an additional nucleotide is inserted from a gene.

base pair the coupling of two complementary bases by hydrogen bonding.

base substitution a form of mutation in which a new base is substituted for a normally occurring nucleotide in a gene.

biolistic process a process in which a projectile is literally shot at a plant cell. The projectile is covered with DNA, which is inserted into the cell as the projectile enters.

biosynthesis the process by which a molecule used in living processes is synthesized in living systems.

blastocyst in early cell division, a spherical structure with a coat, containing numerous dividing cells as well as a fluid-filled cavity.

blastomere a cell found in early cell division after fertilization of an egg.

blunt or flush ends ends of both strands of the DNA that are the same length when DNA is cut using certain restriction endonucleases. See also **sticky ends**.

bond the ability of two atoms to hold together by sharing electrons or by using charge differences or other characteristics of the atoms. Bonds have different strengths, depending on the atoms bonded, the types of bond made, and the solvent in which the atoms are found. See also **covalent bond, disulfide (S-S) bond, hydrogen bond, ionic bond,** and **phosphodiester bond.**

C

calcium sulfate a chemical compound made out of calcium, sulfur, hydrogen, and oxygen, which helps open the pores of cells and allows DNA to enter.

carbohydrate a general name for sugar molecules of all kinds.

catalyst a substance that enhances the rate of a reaction, but is not used up in the process.

catalyze enhance the rate of the reaction.

cDNA complementary single-stranded DNA obtained by making a reverse transcript (DNA from RNA template) from RNA.

cDNA library a collection of bacteria containing fragments of DNA from an organism, which were obtained from reverse transcription of RNA. See also **gene library.**

centrifuge an instrument that spins samples in tubes at high rates of speed; often used to separate biological macromolecules of different sizes.

cellulase an enzyme that attacks the bonds that hold cellulose together; often used to break down the cell walls of plant cells.

chimeric DNA DNA composed of material from two or more sources, such as bacterial/mouse or mouse/human DNA.

chorionic villi sampling sampling of the fluid from the chorionic villi in the placenta.

chromosome a unit of the genetic material of a cell. Some cells, such as bacteria, have a single chromosome. Other cells, such as human cells, contain 23 chromosomes.

cleave or **cleavage** the process by which a chain of amino acids or nucleotides is severed or cut.

clone a cell or organism whose genetic information is identical with that from which it was derived.

codon a three-nucleotide segment of messenger RNA, which uniquely determines insertion of a single amino acid into a protein that will be synthesized.

collagen a three-stranded helical protein of which tendons and cartilage are made.

colony screening method used to determine which bacterial colonies contain certain sequences of DNA.

comb a device that is placed at the top of poured gel to put cavities or wells into the gel so that a sample can be placed in them.

complementarity two strands of nucleic acid that have complementary sequences of nucleotides.

complementary two nucleotides that base pair specifically. In DNA, A-T and G-C are complementary. In RNA, A-U and G-C are complementary.

complementary DNA (cDNA) a strand of DNA complementary to RNA, generally obtained from it using reverse transcriptase.

compound a molecule containing two or more different kinds of atoms.

conjugation that act between bacteria that allows genetic material from one bacterium to be transferred to another.

***cos* sites** sequences of DNA needed to package plasmids into viruses.

cosmid approach method by which cosmids containing plasmids are put into viruses, which in turn transfer the DNA to cells.

cosmids plasmids into which *cos* sites have been inserted, allowing the plasmid to be packaged into a viral coat.

covalent bond a bond between two atoms formed when they share electrons.

crippled virus a virus that cannot reproduce itself inside a host, generally because portions of the viral genome have been removed.

crossing over the act in which DNA from one chromosome exchanges with DNA from another chromosome, allowing the exchange of genetic information.

crown gall tumor a cancer-like growth on a plant that contains bacteria and plant cells, caused by infection with Ti (tumor-inducing) plasmid from *Agrobacterium tumefaciens*.

D

deoxyribonucleotide the fundamental unit found in DNA, containing a deoxyribose sugar, coupled to a phosphate group at the 5' carbon and a nitrogenous base at the 1' carbon.

deoxyribose a five-carbon sugar molecule that contains only -H at the 2' position rather than the –OH group found in ribose; used exclusively in DNA.

differentiation the process in which certain portions of the DNA are used to the exclusion of others, making specialized cells and organelles.

dipeptide two amino acids coupled with a peptide (covalent) bond.

disulfide (S-S) bond a covalent bond between two sulfur atoms, often found in proteins between polypeptide chains.

DNA deoxyribonucleic acid, a long chain of deoxyribonucleotides that contains all the genetic information needed by the organism.

DNA polymerase an enzyme that makes new DNA from deoxyribonucleoside triphosphates using the parent DNA as a template.

DNA profiling sometimes called *fingerprinting*, this technique is used to identify individuals having identical or similar DNA.

double helix the base-paired spiral structure that two strands of DNA ordinarily make.

E

electron an extremely small negatively charged particle, which is located around the nucleus of an atom.

electrophoresis a technique in which samples are moved (pulled) through an acrylamide gel (or some other matrix) using electrical voltage and current.

electroporation the process by which the pores in cell membranes are opened using an electric field; generally used to insert DNA into cells.

element an atom containing a precise number of protons, neutrons, and electrons.

energy the stuff that makes processes go; generally comes in light, heat, mechanical, electrical, or chemical forms.

enzyme a biomolecule, generally a protein, which makes a reaction or process go faster; often used to make or break chemical bonds.

eugenics the science of improving genetic characteristics of the human, which, historically at times, has been accomplished by sterilization or killing people with assumedly negative gene pools.

eukaryote a cell that contains a nucleus, generally more complex than a prokaryote, which contains no nucleus.

eukaryotic a function or process belonging only to eukaryotes.

expressed or **expression** the process by which a gene is expressed as a protein. The DNA makes RNA, which makes protein, the product of the genetic information.

F

frame-shift mutation a deletion or addition of a nucleotide that occurs in the gene, making all the following codons have the wrong three-letter sequence.

G

gamma ray a high-energy beam of very short wavelength.

gene portion of a chromosome or piece of DNA that is used to code for a specific protein.

gene bank a database containing information on the DNA sequence of genes.

gene library a collection of bacteria containing fragments of DNA from an organism, which were obtained by fragmenting the DNA from a cell in that organism. See also **cDNA library**.

genetic code the three-base sequences that specify the order of amino acids to be placed in a polypeptide chain.

genetic linkage the coupling of genetic functions that are close together on the chromosome.

genetics the field of study in which the information contained in the genes is transferred to the offspring.

genome the entire collection of genetic material in a cell, as in the human genome.

germ cell a sperm or egg cell that cannot replicate itself and contains only half the chromosomes of the parent cell.

ghost the protein coat of a virus without the nucleic acid that it normally encapsulates.

glycoprotein a macromolecule that has both a carbohydrate (sugar) portion and a protein, coupled with covalent bonds.

H

HAT medium a growth medium for cells that contains *h*ypoxanthine, *a*minopterin, and *t*hymidine.

helix a spiral form that most nucleic acids take in solution. Two strands intertwined become a double helix, as in DNA.

hemoglobin a large protein, containing four polypeptide chains and carrying up to four oxygen molecules from the lungs to the rest of the organism.

high copy number plasmids that appear in high numbers in some cells.

host a bacterium or a cell that is specifically targeted by a virus.

hybridization the base pairing of DNA with RNA or, more generally, of any nucleic acid strand with any other.

hydrogen bond a weak bond formed between a hydrogen atom and generally an oxygen or a nitrogen atom; very prevalent in proteins and nucleic acids.

hydrophilic water-loving; hydrophilic molecules are easily soluble in water and tend to form hydrogen bonds with water molecules.

hydrophobic water-hating; hydrophobic molecules often avoid water by grouping together and will not form hydrogen bonds.

hydrophobic interactions the grouping together of water-hating molecules that helps to make proteins fold and to stabilize the folded structures.

I

inducer a molecule that helps initiate a specific reaction.

insert a new piece of DNA that is inserted into another piece of DNA.

intron a portion of the gene that is transcribed into RNA, then removed before the RNA is translated.

ion an elemental atom that has either gained or lost one or more electrons and thus has a net charge.

ionic bond a fairly weak bond formed between two oppositely charged groups.

L

lac operon the genomic unit that includes the regulatory regions as well as the structural genes for the lactose region of the genome.

lactose milk sugar composed of two cyclic sugar structures (galactose and glucose).

lambda phage a specific bacteriophage that attacks *Escherichia coli* bacteria.

ligase an enzyme that attaches two pieces of nucleic acid together, forming a phosphodiester bond between them.

linkage forming bonds between molecules.

lipid containing fatty acids or other fat molecules.

lipid bilayer the structure found in cell membranes composed of two layers of lipids, each with their hydrophilic groups pointing out and the hydrophobic lipid portions pointing into the membrane.

lipoprotein a lipid (fat) and a protein attached together by a covalent bond; often found in cell membranes.

liposome carrier a small vesicle made of lipid bilayers, such as a cell membrane that can carry nucleic acids or other molecules into cells.

long chains amino acids linked together in proteins (polypeptide chains); in nucleic acids, made of nucleotides linked together.

lysis breaking cell membranes open; done naturally to release viruses after they have reproduced themselves within the cells.

lysogenic pathway after a bacteriophage infects a bacterium, the DNA from the virus is inserted into the bacterial chromosome and remains there until a lysogenic event occurs (often stimulation with ultraviolet light), at which time the viral DNA begins to make new viruses.

lytic pathway after a bacteriophage infects a bacterium, the viral DNA is immediately used to make new viruses, which are released upon lysis of the cell membrane.

M

macromolecule a large molecule composed of many smaller molecules and compounds.

marker a substance used to see proteins or nucleic acids, as a stain or radioactivity on a gel. A marker is also a portion of DNA that can be identified for its function or nonfunction. Antibiotic resistance is a marker.

minichromosome another term for a *plasmid*.

missense mutation any mutation producing a misreading of a codon.

mitochondrion the part of a cell that transforms energy for the cell; the powerhouse.

molecule a substance containing two or more atoms.

mRNA messenger RNA; the RNA that carries the genetic message to the ribosome.

multilocus analysis typing DNA using probes to several alleles or VNTR or STR regions.

mutagen anything that causes a mutation to take place, such as ultraviolet light.

mutant a cell or organism containing mutations in the DNA. The results of the mutation are often seen as some small or large transformation of cell or organism structure or function.

mutation the substitution, deletion, or insertion of a nucleotide in genomic DNA.

myoblast precursor of skeletal muscle cells.

N

negative control an experimental control designed to show that the experiment is not working when it should not work.

negative strand the strand of DNA used as a template to make a complementary (positive) strand of RNA to be used as a message.

neutral mutation mutation that causes no apparent effect on protein structure or activity.

novelty legal term used in patent law suggesting that the invention is new and different from any other invention. See also **obviousness** and **utility**.

nucleic acids long chains of nucleotides coupled by phosphodiester bonds. There are two types, deoxyribonucleic acid (DNA) and ribonucleic acid (RNA).

nuclein the early name given to the material extracted from the nuclei of cells.

nucleosome structures in which about 200 base pairs of DNA are wrapped around proteins (histones). Nucleosomes occur all along a strand of DNA giving it a beaded structure.

nucleotide a macromolecule containing a sugar (either ribose or deoxyribose), a phosphate group, and a nitrogenous base (see fig. 1–12).

nucleus a portion of a eukaryotic cell that is partitioned by a nuclear membrane and contains the genetic information (DNA) of the cell.

O

obviousness legal term used in patent law to identify new ideas. If the invention would be obvious to a person working in the field, the idea is not patentable. See also **novelty** and **utility.**

OH group hydroxyl group, often attached to carbon and other atoms. This little group helps molecules become soluble in water.

operon the name of an entire regulated region of DNA, consisting of a regulator region, a control region, and the codes for the proteins to be made.

organelle literally a small organ; applied to small functional units within cells, generally.

organism a complete, living body—either one-celled such as bacteria or multicelled such as plants and animals.

P

packaging as applied in this book, the act of putting the necessary parts into a virus. See also **packaging cell.**

packaging cell a cell line that is able to provide the missing pieces to allow viruses containing inserts to be packaged into functional units.

patent a legal assignment of exclusive rights to market an invention to the inventor for a period of 20 years.

PCR see **polymerase chain reaction**.

peptide bond a special name given to the covalent bond between the amino group (NH) of one amino acid with the carbonyl group (C=O) of the adjacent amino acid.

Petri dish flat glass dish about 3 inches in diameter with short vertical sides; generally filled with about $1/4$ inch of agar upon which bacteria are spread and grow.

phage bacteriophage, a bacterial virus.

phosphodiester bond the bond containing phosphorous which occurs between the 3' carbon of one nucleotide and the 5' carbon of the adjacent nucleotide in a nucleic acid.

phosphorylation the act of placing a phosphate (PO_3) group on another molecule.

pilus an extension protruding from a bacterium by which it couples with another bacteria and transfers genetic information.

plasmid a minichromosome; a small piece of DNA, normally circular, which is commonly found in bacteria and contains sex factors, antibiotic resistance genes, and other material.

plasmid transfer the act of transferring a plasmid from one organism to another.

pluripotent descriptive of cells that are not differentiated and can use all of their genetic material. Same as **totipotent**.

point mutation a mutation that occurs at a single base pair within the DNA.

polyacrylamide gel polymerized acrylamide, a gelatin-like substance used in gel electrophoresis.

polymerase chain reaction (PCR) process in which the amount of DNA is amplified by making multiple copies of the two strands.

polymerization making long chains of molecules out of fundamental building blocks.

polymorphic variations that occur at specific allelic regions in the DNA causing differences in restriction fragments.

polypeptide chain chain of amino acids linked by peptide bonds, generally containing between about 20 and 100 amino acids.

positive control a regulatory action that activates transcription.

positive strand in double-stranded DNA, the positive strand is the one that contains the genetic message.

precipitate the substance that comes out of solution upon adding certain chemicals to the solution; can generally be seen in the light, making the solution cloudy.

primary structure the initial sequence of amino acids or nucleotides in the polypeptide chain or nucleic acid, respectively.

primer a short piece of DNA (or RNA) that is used by DNA or RNA polymerase to begin the process of replication or transcription or reverse transcription.

probe a general term referring to anything that can bind a specific region of nucleic acid or protein; generally a short piece of DNA complementary to a region of DNA or RNA.

prokaryotes cells that do not contain nuclei.

promoter a region of the DNA that binds RNA polymerase to initiate transcription.

pronucleus the unfertilized nucleus in an egg cell or a sperm cell. Upon fertilization, two pronuclei may exist for a short period of time in the egg cell—one from the egg and one from the sperm.

protein one or more polypeptide strands folded in a specific manner to allow biological function.

proton a positively charged particle within the nucleus of an atom.

protoplast a plant cell from which the cell wall has been removed.

R

recognition site a site in the DNA containing a specific sequence of bases that are recognized by a certain restriction endonuclease.

recombination the process by which a different double-stranded segment of DNA combines with the original chromosomal DNA, inserting the new DNA into the chromosome.

regulation a general term dealing with the control of DNA transcription and the resultant expression of proteins through translation.

regulator regions regions of DNA that are used to control the activity of transcription and other control functions.

replica plating making an exact duplicate of the colonies on a Petri dish by using a piece of sterile felt (or another similar device) to transfer portions of all the colonies from one dish to another.

replication the process of making new DNA from parent DNA.

repressor a protein that is used to turn off the transcription process by binding to the control region of DNA, thereby not allowing the RNA polymerase to initiate transcription.

restriction endonuclease an enzyme that binds to a specific region of DNA molecules and cleaves (splits) them at a specific site within or adjacent to the specific binding region; often called a *restriction enzyme*.

restriction fragment the pieces of DNA that result when DNA is cleaved with a restriction endonuclease.

restriction fragment length polymorphism (RFLP) length variations in DNA fragments that result from restriction enzyme cleavage of specific allelic regions in the DNA from different individuals with allelic variations. RFLPs are visualized on an acrylamide gel and show DNA fragment length variations between individuals.

restrictive environment a chamber or room designed specifically to restrict the entrance or exit of any new material. Such a facility is used to work with highly toxic chemicals or viruses, such as HIV, which are extremely hazardous.

retrovirus the general name for viruses that contain RNA as their genome.

reverse transcriptase an enzyme that makes DNA from an RNA template.

reversion a mutational event that either reverses the original mutation or compensates for it at another location in the genome.

ribose a five-carbon sugar that is used to make nucleic acids.

ribosomal RNA (rRNA) the RNA found in ribosomes.

RNA ribonucleic acid, a long chain of ribonucleotides covalently bound together. See also **messenger RNA (mRNA), transfer RNA (tRNA), and ribosomal RNA (rRNA).**

S

sandwich assay a name applied to an assay that uses a DNA probe to target a DNA sequence in an unknown piece of DNA, after which a labeled (with a fluorescent or radioactive label) probe is bound to another region of the original DNA probe.

screening a technique used to determine whether a certain DNA sequence is present or not. See also **colony screening.**

secondary structure the helical twisting or bending and turning of the chain of amino acids or nucleotides.

short tandem repeat (STR) short sequences (generally 3 to 7 nucleotides in length) that are repeated multiple times adjacent to each other, flanked by uniform regions of DNA.

sickle cell crescent-shaped red blood cells that contain mutant (sickle cell) hemoglobin.

sickle cell anemia the disease caused by the presence of sickle cell hemoglobin in the red blood cells.

side chain the term applied to the different functional groups occurring on the alpha carbon of amino acids.

silent mutation mutation in the gene that does not alter the function of the resultant protein.

single locus analysis typing DNA using a probe to a single allele or a single VNTR or STR region.

somatic cells all the cells of the organism except the germ (sperm and egg) cells.

Southern blot the technique developed by Dr. Southern in which DNA fragments in agarose gels are transferred to nitrocellulose paper by pulling them out of the gel using absorbent, blotting material to wick the solution through the gel.

start codon the first three-letter code in the string of codons that code for a protein; ordinarily AUG.

stem cell the pluripotent cell from which all other cells are made.

sticky end an overlapping end on a strand of DNA complementary to the overlapping end on the other strand of DNA; found with restriction fragments.

stop codon the codon that signals the termination of translation of a peptide chain; generally UAA, UAG, or UGA.

synthesize the act of making new products, either biologically or chemically.

T

Taq **polymerase** a DNA polymerase (an enzyme that makes DNA) derived from *Thermus aquaticus*, an organism that grows in high temperature surroundings (found in Yellowstone Park). This enzyme is used for the PCR process.

template the strand of DNA or RNA used as a pattern by DNA or RNA polymerase from which to make a complementary copy.

tertiary structure the result of the folding of helical turns or other turns of polypeptide chains into a function protein.

thymidine kinase an enzyme that adds a phosphate group to thymidine, which is necessary for the manufacture of DNA.

Ti plasmid the tumor-inducing plasmid found in *Agrobacterium tumefaciens*, which can be inserted into plant cells.

tissue culture the process in which various eukaryotic cell lines are grown outside the whole organism, much as bacteria are grown.

totipotent descriptive of cells that are not differentiated and can use all of their genetic material. Same as **pluripotent**.

transduction transfer of a bacterial gene from one bacterium to another by means of a bacteriophage.

transcription making RNA, using RNA polymerase, using DNA as a template.

transcriptional control the regulation of transcription by various means to limit or enhance the amount of proteins produced.

transfection the addition of new genetic material to eukaryotic cells. Same as **transformation** of bacterial cells.

transformation the addition of new genetic material to bacterial cells. Same as **transfection** in eukaryotic cells.

transgenic animal animal containing new genetic material in its germ cells.

translation the process in which polypeptide chains are made according the coded information in the messenger RNA (mRNA).

tRNA transfer RNA; a small RNA molecule that carries an amino acid to the ribosome and places it in the proper position in the polypeptide chain that is being formed.

tumor-infiltrating lymphocyte (TIL) a cell line that has been genetically modified, which seeks out tumor cells.

U

ultraviolet (UV) light light having a wavelength shorter than is visible, which can cause damage to DNA.

utility one of the legal terms used in patent law suggesting that the invention has a use that is new or different from any other invention. See also **novelty** and **obviousness**.

V

variable number tandem repeat (VNTR) sequence of DNA that is repeated multiple times adjacent to each other, having identical flanking sequences in all humans. However, the number of repeats varies, depending on the individual.

vector a plasmid or a phage that is used to carry new genetic material into a cell.

viral plaque a region on a bacterial lawn in which the bacteria have been lysed owing to the presence of phage.

virus a vector that generally contains a protein coat with a nucleic acid genome inside. See also **bacteriophage.**

W

wells the regions in an acrylamide gels into which the sample is placed before electrophoresis begins.

Z

zygote a fertilized egg.

Appendix
The amino acids

Name	Abbreviation		Structure
	3 letters,	1 letter	

Alanine Ala A

$$CH_3$$
$$H_3N^+\!-\!C\!-\!COO^-$$
$$H$$

Arginine Arg R

$$NH_2$$
$$C=NH_2^+$$
$$NH$$
$$CH_2$$
$$CH_2$$
$$CH_2$$
$$H_3N^+\!-\!C\!-\!COO^-$$
$$H$$

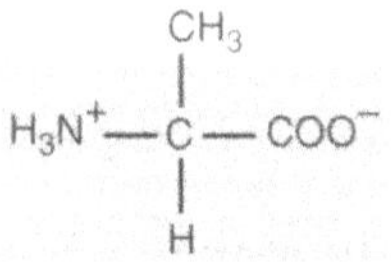

Asparagine Asn N

$$O=C\diagup NH_2$$
$$CH_2$$
$$H_3N^+\!-\!C\!-\!COO^-$$
$$H$$

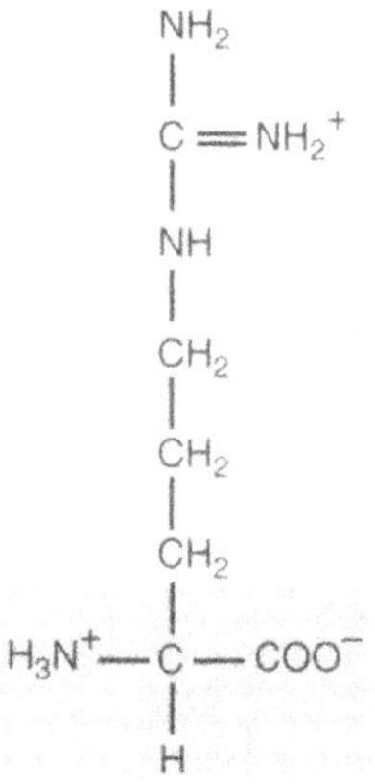

Aspartic acid Asp D

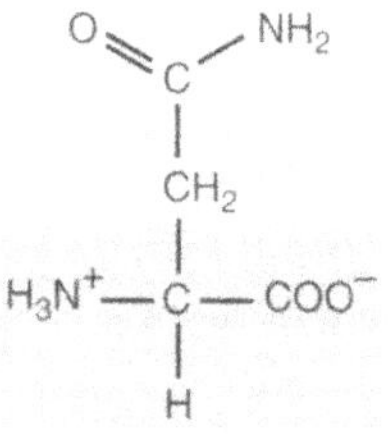

$$O=C\diagup O^-$$
$$CH_2$$
$$H_3N^+\!-\!C\!-\!COO^-$$
$$H$$

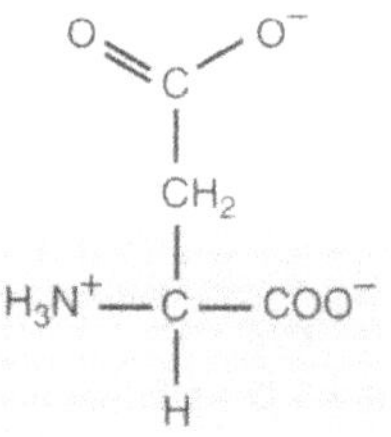

Cysteine Cys C

$$SH_2$$
$$CH_2$$
$$H_3N^+\!-\!C\!-\!COO^-$$
$$H$$

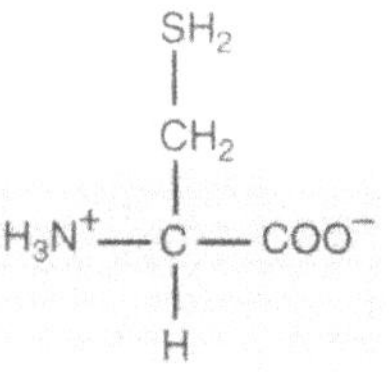

Name	Abbreviation 3 letters, 1 letter	Structure
Glutamine	Gln Q	$O=C-NH_2$ $\vert$ CH_2 $\vert$ CH_2 $\vert$ $H_3N^+—C—COO^-$ $\vert$ H
Glutamic acid	Glu E	$O=C-O^-$ $\vert$ CH_2 $\vert$ CH_2 $\vert$ $H_3N^+—C—COO^-$ $\vert$ H
Glycine	Gly G	H $\vert$ $H_3N^+—C—COO^-$ $\vert$ H
Histidine	His H	$HN{=}\!{\diagup}\!N^+H$ (imidazole ring) $\vert$ CH_2 $\vert$ $H_3N^+—C—COO^-$ $\vert$ H
Isoleucine	Ile I	CH_3 $\vert$ CH_2 $\vert$ $CH_3—CH$ $\vert$ $H_3N^+—C—COO^-$ $\vert$ H

Name	Abbreviation 3 letters,	1 letter	Structure
Leucine	Leu	L	H_3C–CH–CH_3 / CH_2 / H_3N^+–C–COO^- / H
Lysine	Lys	K	NH_3^+ / CH_2 / CH_2 / CH_2 / CH_2 / H_3N^+–C–COO^- / H
Methionine	Met	M	CH_3 / S / CH_2 / CH_2 / H_3N^+–C–COO^- / H
Phenylalanine	Phe	F	CH_2 / H_3N^+–C–COO^- / H
Proline	Pro	P	CH_2 / CH_2 CH_2 / H_2N^+–C–COO^- / H

Name	Abbreviation 3 letters, 1 letter	Structure
Serine	Ser	S	 OH | CH$_2$ | H$_3$N$^+$—C—COO$^-$ | H
Threonine	Thr	T	 CH$_3$ | HCOH | H$_3$N$^+$—C—COO$^-$ | H
Tryptophan	Trp	W	(indole ring) CH$_2$ | H$_3$N$^+$—C—COO$^-$ | H
Tyrosine	Tyr	Y	OH (benzene ring) CH$_2$ | H$_3$N$^+$—C—COO$^-$ | H
Valine	Val	V	H$_3$C CH$_3$ CH | H$_3$N$^+$—C—COO$^-$ | H

INDEX

ABOUT THE AUTHOR

Walter E. Hill received his undergraduate education at Pomona College and Brigham Young University, graduating from the latter in 1961 with a B.S. Degree in Physics. Graduate work at the University of Wisconsin centered on the physical structure of ribosomes, the protein factories of all cells. Graduating from Wisconsin in 1967 with a Ph.D. in Biophysics, Dr. Hill continued his studies as an NIH Post-Doctoral Associate under the direction of Dr. Kensal Van Holde, at Oregon State University, using the techniques of physical biochemistry of further study ribosome structure. As an Assistant Professor at the University of Montana, Dr. Hill was awarded an *NIH Career Development Award* to continue his studies of ribosome structure and function. These studies have been continued with support from the National Science Foundation and the National Institutes of Health over the last thirty years. During this period Dr. Hill has received numerous awards, including the *Distinguished Scholar Award* from the University of Montana, the *Burlington Norther Distinguished Researcher Award* and the *Montana Academy of Sciences Mershon Award* for Distinguished Research. During this period he has served on many review panels and study sections, reviewed numerous manuscripts and grants, and published over 50 scholarly articles in scientific journals, written nine book chapters and edited two books on ribosomes.

9 780367 454906